Ecological Studies, Vol. 74

Analysis and Synthesis

Ecological Studies

Volume 61
Potentials and Limitations of Ecosystem Analysis
Edited by E.-D. Schulze and H. Zwolfer
1987. XII, 435p., 141 figures. cloth
ISBN 0-387-17138-X

Volume 62
Frost Survival of Plants
By A. Sakai and W. Larcher
1987. XII, 321p., 200 figures, 78 tables. cloth
ISBN 0-387-17332-3

Volume 63
Long-Term Forest Dynamics of the Temperate Zone
By Paul A. Delcourt and Hazel R. Delcourt
1987. XIV, 450p., 90 figures, 333 maps. cloth
ISBN 0-387-96495-9

Volume 64
Landscape Heterogeneity and Disturbance
Edited by Monica Goigel Turner
1987. XII, 241p., 56 figures. cloth
ISBN 0-387-96497-5

Volume 65
Community Ecology of Sea Otters
Edited by G.R. van Blaricom and J.A. Estes
1987. X, 280p., 71 figures. cloth
ISBN 3-540-18090-7

Volume 66
Forest Hydrology and Ecology at Coweeta
Edited by W.T. Swank and D.A. Crossley, Jr.
1987. XIV, 512p., 151 figures. cloth
ISBN 0-387-96547-5

Volume 67
Concepts of Ecosystem Ecology A Comparative View
Edited by L.R. Pomeroy and J.J. Alberts
1988. XII, 384p., 93 figures. cloth
ISBN 0-387-96686-2

Volume 68
Stable Isotopes in Ecological Research
Edited by P.W. Rundel, J.R. Ehleringer and K.A. Nagy
1989. XVI, 544p., 164 figures. cloth
ISBN 0-387-96712-5

Volume 69
Vertebrates in Complex Tropical Systems
Edited by M.L. Harmelin-Vivien and F. Bourliere
1989. XI, 200p., 17 figures. cloth
ISBN 0-387-96740-0

Volume 70
The Northern Forest Border in Canada and Alaska
By James A. Larsen
1989. XVI, 272p., 73 figures. cloth
ISBN 0-387-96753-2

Volume 71
Tidal Flat Estuaries: Simulation and Analysis of the Ems Estuary
Edited by J. Baretta and P. Ruardij
1988. XVII, 353p., 1234 figures. cloth
ISBN 3-540-19323-5

Volume 72
Acidic Deposition and Forest Soils
By D. Binkley, C.T. Driscoll, H.L. Allen, P. Schoeneberger, and D. McAvoy
1989. 152p., 34 figures. cloth
ISBN 0-387-96889-X

Volume 73
Toxic Organic Chemicals in Porous Media
Edited by Z. Gerstl, Y. Chen, U. Mingelgrin, and B. Yaron
1989. XIV, 343p., 88 figures. cloth
ISBN 3-540-50799-X

Volume 74
Inorganic Contaminants in the Vadose Zone
Edited by B. Bar-Yosef, N.J. Barrow and J. Goldshmid
1989. XIII, 200p., 70 figures. cloth
ISBN 3-540-50788-4

B. Bar-Yosef N. J. Barrow J. Goldshmid (Eds.)

Inorganic Contaminants in the Vadose Zone

With 70 Figures

Springer-Verlag
Berlin Heidelberg New York
London Paris Tokyo Hong Kong

Dr. Bnayahu Bar-Yosef
Agricultural Research Organization
The Volcani Center
Institute of Soils and Water
P.O. Box 6
Bet Dagan 50250, Israel

Dr. N. J. Barrow
CSIRO Division of Animal Production
Private Bag
PO Wembley 6014
Western Australia, Australia

Dr. J. Goldshmid
Environmental Engineering & Design Comp. Ltd.
P.O. Box 43059
Tel Aviv 61430, Israel

ISBN-13:978-3-642-74453-2 e-ISBN-13:978-3-642-74451-8
DOI: 10.1007/978-3-642-74451-8

Library of Congress Cataloging-in-Publication Data. Inorganic contaminants in the vadose zone / B. Bar-Yosef, N. J. Barrow, and J. Goldshmid, eds. p. cm. – (Ecological studies; vol. 74) Selected papers from an international workshop hosted by the Institute of Soils and Water of the Agricultural Research Organization, Israel, June 1987, sponsored by the International Union of Pure and Applied Chemistry and International Association of Hydrological Sciences. Includes index.ISBN-13:978-3-642-74453-2(U.S.).Inorganic compounds–Environmental aspects – Congresses. 2. Soil pollution – Congresses. 3. Zone of aeration – Congresses. I. Bar-Yosef, B. (Bnayahu), 1941– II. Barrow, N. J. III. Goldshmid, J. (Jhuda), 1931– . IV. Makhon le-kạrkạʻ u-mayim (Israel) V. International Union of Pure and Applied Chemistry. VI. International Association of Hydrological Sciences. VII. Series: Ecological Studies; v. 74. TD879.I55156 1989 574.5'26404 – dc 20

Softcover reprint of the hardcover 1st edition 1989

Typesetting: International Typesetters Inc., Makati, Philippines
2131/3145-543210 – Printed on acid-free paper

Preface

The understanding of pollutant transformations, sorption and transport in soils and uptake by plants is the key to controlling contaminant movement towards groundwater and accumulation in food-chains. Some of these major processes were discussed in the first International Workshop on Pollutants in Porous Media, convened in Israel in 1983 (Pollutants in Porous Media, Eds. Yaron, Dagan and Goldshmid, Springer-Verlag, 1984). Since that meeting, research on contaminant impact on the environment expanded considerably. New mechanisms explaining inorganic pollutant partitioning between solid and solution phases in soils were proposed and tested; specific elements and specific reactions were investigated, and more case studies to evaluate management practice effects on drinking water quality and accumulation of contaminants in crops were carried out. The recognition that the new knowledge needs to be discussed and evaluated, and the scientific and engineering communities be updated, led the IUPAC (International Union of Pure and Applied Chemistry) and IAHS (International Association of Hydrological Sciences) Water Quality Commissions to organize a second International Workshop on pollutant behaviour in the vadose zone, hosted by the Institute of Soils and Water of the Agricultural Research Organization, Israel. In this workshop, taking place in June 1987, attention was focused on chemical, physico-chemical and microbial-mediated reactions of different contaminants in the vadose zone. This book contains selected works presented in the 1987 workshop by invited specialists from various disciplines and countries. It is supplemented by the proceedings of the other part of the 1987 workshop, which dealt with organic contaminants in the vadose zone (Eds. Gerstl, Chen, Mingelgrin, and Yaron, Springer-Verlag, 1989).

The contributions in this book were grouped into four parts: (1) Processes and Principles of Inorganic Pollutant Behaviour in Reactive Porous Media, with emphasis on modelling ion partitioning between solution and solid phases, and the effect of solute movement on microbial transformations in soils. (2) Behaviour of Specific Elements in the Soil, and its dependence on pH, redox potential and microbial activity. (3) Management of Inorganic

Pollutants, with emphasis on pollutant discharge rate from various waste sources, uptake by crops and effects on soil properties. (4) Case Studies, showing effects of improved fertilization management on pollutant load in soil, and a method to trace ion transport to deep underground water under field conditions.

We acknowledge the support given to the workshop by the Israeli Ministry of Agriculture, Ministry of Science and Development, the Seagram Foundation for Soil and Water Sciences of the Hebrew University, and the US-Israel Binational Agricultural Research and Development Fund. We thank the workshop participants for their written contributions, and Springer-Verlag for undertaking the publication of this book.

The Editors

Contents

Part C Management of Inorganic Pollutants

Part D Case Studies

List of Contributors

You will find the addresses at the beginning of the respective contribution

Part A Processes and Principles of Pollutants Behaviour in Reactive Porous Media

Introductory Comments. The fate of waste materials and fertilizers deposited in soils depends on complex interaction and physical, chemical and biological processes. A major gap in current understanding of pollutant fate in soils is the ability to predict ion partitioning between solid and solution phases of reactive porous media, especially in the presence of complexing organic compounds. Traditional models, reviewed in Barrow's contribution, yield acceptable estimates of cation and anion sorption as a function of the ion solution concentration and pH under equilibrium or quasi-equilibrium conditions. Those models assume that the adsorbing surface is uniform and unchanging with time. Kafkafi reviewed recent evidence indicating that face-to-edge interaction may play an important role in determining ion adsorption by clay minerals predominant in soils. Robert and Terce presented data showing that clay-particle organization and Fe and Al coatings should be considered when modelling pollutant sorption by porous media. Chambers et al. discussed the importance of organo-metal complexes in determining pollutant mobility, and provided data allowing a better incorporation of the formed complexes into sorption models. The effect of time on adsorption, and the importance of including adsorption kinetics in ion transport models was discussed by Barrow.

The role of microbial transformations in determining the fate of contaminants in soils is discussed in the following chapters. However, Darrah's contribution in this part deals with the principles of micro-scale solute movement effects on mineralization reactions, thus showing that ion partitioning and microbial-mediated processes in soils are closely interrelated.

B. Bar-Yosef

Organo-Metal Complexes as Mobility-Determining Factors of Inorganic Toxic Elements in Porous Media

H. Klamberg[1], G. Matthess[2], and A. Pekdeger[3]

1 Introduction

The transport behaviour of toxic metals in the subsurface is controlled by hydraulic processes and by the interactions between water and solid rock and soil materials by dissolution-precipitation, acid-base reactions, oxidation-reduction, complexation and adsorption-desorption. These processes, which have been discussed in a previous contribution (Matthess 1984a), are affected by natural organic substances. In this chapter only organo-complex formation, and sorption and desorption of toxic metals at solid organic substance affecting the mobility of toxic metals are considered.

In concentrated aqueous solutions (for ionic strengths above 0.1 M, but in many cases also for much lower values) complex ions appear in solution. These can be treated as thermodynamic entities and are in dynamic equilibrium with the free ions. Complexes are combinations of cations called central atoms with molecules or anions, containing free pairs of electrons (bases), referred to as ligands. Bases containing more than one ligand atom, e.g. oxalate and citrate, are referred to as multidentate complex formers. Complex formation with multidentate ligands is called chelation, and the complexes are called chelates. Chelates are usually more stable than complexes with monodentate ligands (Garrels and Christ 1965; Stumm and Morgan 1981). As a measure of the stability of a complex, a stability constant can be defined. The complex stability constant of the complexes in the reaction (1):

$$mM + nL = M_mL_n \tag{1}$$

is defined by Eq. (2)

$$K = \frac{[M_mL_n]}{[M]^m[L]^n}. \tag{2}$$

Complexation may change the solubility. In pore solutions of the unsaturated zone fulvic acids, tartaric acid, citric acid and salicylic acid tend to increase cation solubility. Thus metals, especially iron, are transported in connection with naturally dissolved organic substances (Flaig et al. 1975). In soil solutions Fe^{2+}

[1]Department of Chemistry, Marburg University, Hans-Meerwein-Str., D-3550 Marburg, FRG
[2]Institute of Geology and Paleontology, Kiel University, Olshausenstr. 40–60, D-2300 Kiel, FRG
[3]Institute of Applied Geology, Free University, Wichernstr. 16, 1000 Berlin, 33, FRG

seems to be present mainly as organic complexes and to be removed from the seepage by the oxidation and precipitation of the dissolved organic substances (Matthess et al. 1977).

Humic acids form metal ion chelates of low solubility occupying two-thirds of the total binding capacity of humic acids, which is about 200–600 mEq metal ion/100 g (Förstner and Müller 1974).

For the numerous complex-forming substances in the soil, a general decrease of complex stability can be stated (Scheffer & Schachtschabel, 1982):

$$Fe^{3+} > Al^{3+} > Cu^{2+} > Pb^{2+} > Fe^{2+} > Ni^{2+} > Cd^{2+} > Zn^{2+} > Mn^{2+}.$$

Humic acid complexes of Cu, Pb, Cd and Zn have log K-values of 8.65, 8.35, 6.25 and 5.72, whereas for fulvic acid complexes of Cu, Pb and Zn log K-values of 4.0, 4.0 and 3.6 respectively are quoted (Scheffer and Schachtschabel 1982).

Dissolved metals may be sorbed by soil and rock materials, among which, besides clay minerals, humic substances are the most effective. The equilibrium between dissolved and suspended substances is very often described by the Henry isotherm, which defines the equilibrium between the concentrations of the dissolved (C_s) and adsorbed (C_a) species in dilute solutions:

$$C_a = K_d \cdot C_s. \tag{3}$$

The distribution coefficient K_d is widely used as a parameter to describe the affinity of a specific geological material for a certain substance and the geochemical transport behaviour. It is defined as the ratio of the adsorbed substance in mEq g^{-1} solid to the concentration of the same substance in solution in mEq l^{-1}. It is linked to the retardation factor R_d, the quotient of the mean groundwater velocity (v_w) to the mean transport velocity of the dissolved substance (v_c), e.g. in the widely used one-dimensional retardation equation:

$$R_d = \frac{v_w}{v_c} = 1 + \frac{\delta_b}{n} \cdot K_d, \tag{4}$$

using the bulk density of the aquifer material (δ_b) and its porosity (n). The retardation is specific for any dissolved or suspended substance with respect to the surrounding groundwater. Thus, a chromatography effect can be observed. This approach holds only if one reversible, kinetically fast adsorption process is present and no changes in the speciation of dissolved and solid partners occur (Jackson 1980). If these boundary conditions are not given, the transport characteristics of a contaminant deviate from the presented model behaviour. Batch and column experiments with sand materials and solutions of Cd, Zn, Cu and Tl did not allow a correct prediction of their transport velocity, since complexing, precipitation and other processes do not establish the linear relationship and reversibility necessary for the isotherm concept (Isenbeck et al. 1985).

Another description of sorption-desorption equilibria may be the Langmuir isotherm (5), which is possibly a better mathematical definition:

$$c_a = \frac{k_1 \cdot k_2 \cdot C_s}{1 + k_2 \cdot C_s}, \tag{5}$$

where k_1 is the adsorption maximum when the adsorbent is completely saturated and k_2 is a constant relating to the bonding energy (sorption binding constant).

The hydrogeological implication of the complexing of the transition metal ions (e.g. Fe, Mn, Co, Zn, Pb, Cu, Ni, Cr) by inorganic or organic ligands is to alter their adsorption or precipitation from solution. For example, the adsorption of radioactive cobalt and zinc is significantly decreased in the presence of dissolved organic carbon (DOC), which is found in all natural waters. This decrease is due to the formation of organo-cobalt and organo-zinc complexes whose sorption behaviour may differ from that of the hydrated metal ions and whose increased solubility in solution reduces the effectiveness of precipitation (Jackson et al. 1980).

2 Organic Substances in Natural Porous Media

Natural and man-made organic substances are present in the subsurface. The natural organic compounds are humic substances as major constituents and minor amounts of e.g. simple fatty acids, tannic acid, amino acids and various degradation products. They originate either from the biologically active soil zone or from aquifer materials. The humic substances can be divided according to their behaviour in aqueous solutions at certain pH values into the fulvic acids to the humins, however, the molar mass increases. Fulvic acids comprise numerous heterogeneous compounds that differ from humic acids and humins by having a lower molar mass and a higher content of oxygen-containing functional groups (-COOH, -OH, > CO). Humic acids are mostly highly polymerized spherocolloids. The chemical structure and properties of the humins seem to be similar to the humic acids, however, they are firmly adsorbed or bonded to soil and sediment constituents (Schnitzer and Khan 1978; Scheffer and Schachtschabel 1982).

The highest organic C contents are found in peat, lignite, coal, and in oil- and gas-bearing deposits. High organic C contents are present in the top soil, depending of the soil type: 0.05–4.30 wt%. The mean contents of organic carbon vary according to the rock type in shales between 0.40 and 1.68 wt%, in carbonates between 0.15 and 0.50 wt%, in sandstones between 0.04 and 0.42 wt% and in quaternary sands and gravels between 0.008 and 1.9 wt% (Wedepohl 1978; Matthess 1984b).

The dissolved organic substances are mostly determined as total organic carbon (TOC) or dissolved organic carbon (DOC). In natural groundwaters TOC usually is in the order of a few mg l^{-1}, but in groundwater which is in contact with oil or peat much higher concentrations are observed, e.g. in the order of hundreds of mg l^{-1} (Matthess 1984b). The content of dissolved organic substances is highest in the seepage water of the uppermost top layers of the soils rich in organic material. It decreases more or less rapidly towards the mineral soil horizons. The highest contents (up to 2000 mg l^{-1} permanganate demand) have been observed in the seepage water of the crude humus horizon of the podzols (Matthess and Pekdeger 1980). The decrease of dissolved organic substances with depth may be due to adsorption of the organic substances onto clay minerals or to forming organo-mineral compounds or due to precipitation, flocculation and forming of unsoluble complexes. Finally, the dissolved organic substances can be degraded by microbial oxidation (Schulz 1970). The first two processes prevail relatively in the upper soil horizons, whereas degradation processes gain importance with growing depth. This explains the small amount of organic substances leached from the soil, which can be found as well in the pore solutions as in the solid phase of deeper layers. The changes of the composition of the dissolved organic substances, especially the humic substances, can be observed by fractionating them with the help of molecular sieves following Flaig et al. (1975) and Schnitzer and Khan (1978). The subsequent analysis steps use, e.g. infrared spectrometry or mass spectrometry-gas chromatography.

By combined use of gel filtration (Sephadex G-25) and extinction measurements in the seepage of podzol and brown forest soils, Matthess and Pekdeger (1980) could show that high molecular humic and fulvic acids are major components of the organic content of seepage water.

The absolute contents of organic substances in the seepage of different depths, shown from a podzol (Pekdeger 1977; Matthess et al. 1977), decrease without substantial changes in their molecular size spectrum. Average molecular sizes seem to be the most resistant, whereas large molecules are predominantly adsorbed or precipitated, small molecules are preferentially degraded.

The following experiments aimed at obtaining a better understanding of the influence of organic complexes on toxic metal transport in a podzolic soil, which is a widespread soil type in the northern parts of Europe, Asia and North America. Furthermore, the measurements should provide better stability constants for geochemical model calculations of rock-water interaction in the investigated system using programs like WATEQF (Plummer et al. 1976) or PHREEQE (Parkhurst et al. 1980).

3 Experimental Procedures

3.1 Origin and Properties of Samples

The monolith lysimeters and the humic substances from the A_h- and B_h-horizons used for the following experiments were taken from a podzol site on Quaternary fluvio-glacial outwash sands and gravel in the Segeberger Forest north of Hamburg, West Germany. The humic substances were isolated by extraction of the soil with NaOH (0.1 mol l^{-1}), precipitation with HCl at pH 1 and subsequent cleaning procedures. These materials were characterized in a general manner by determination of the C, H, N, water and ash content. The results evaluated for water-free (wf) and water- and ash-free (waf) substances are listed in Table 1. The nitrogen content increases with depth. The Fe contents were measured after treatment with $HClO_4/HNO_3$ and evaluated for ash-free substances. These values do not correlate with the ash content of the samples. The E_4/E_6-ratios, i.e. the quotient of the extinctions of monochromatic light at 465 and 665 nm (Chen et al. 1977), were determined from the extinctions of gel-chromatographic fractions of humic acid solutions in NaOH (Post 1987).

Table 1. Composition of soil humic acids (waf = free of water and ash, wf = free of water)

Horizon	Depth (cm)	(waf) C (%)	H (%)	N (%)	(wf) Fe (%)	Ash (%)	E_4/E_6
Ah_e	11–25	57.59	4.65	2.83	0.18	5.7	5.45
A_e	25–45	60.42	4.85	2.68	0.24	20.4	5.53
$A(s)_h$	45–52	58.15	5.14	3.19	0.44	7.4	5.90
$B(h)_s$	52–68	58.32	5.43	3.88	0.37	8.1	6.84

The infrared spectra of all samples follow the typcial pattern of humic acids, which indicates the presence of different functional groups (-OH at 3400 cm^{-1}, -COOH at 2500–3200 cm^{-1}, C=O at 1720 cm^{-1}), aromatic (3080 cm^{-1}) and aliphatic (2920 and 2850 cm^{-1}). The extinction of UV/VIS spectra shows the usual monotonous decrease with increasing wavelength. The E_4/E_6 ratio increases with depth under simultaneous decrease of total extinction, which according to Chen et al. (1977) indicates a decreasing molecular size. The molar masses of the materials were determined with vapour pressure osmometry and gel permeation chromatography with Fractogel TSK HW-40, HW-50 and HW-55. As fluid, a Tris-buffer solution (0.01 mol l^{-1}) with an addition of KCl (0.1 mol l^{-1}) was used to keep the irreversible, adsorbed part below about 10%. The mean molar masses of the soil humic acids are according to the vapour pressure osmometric measurements about 1900 (A_h) and about 1400 (B_h), whereas the gel-chromatographic measurements indicate molar masses in the order of 5000–20 000. Mass spectrometric measurements support the higher gel-chromatographic results.

3.2 Monolith Lysimeter Experiments

Lysimeters of various lengths (0.1–1 m), involving different soil horizons ($O + A_h$, $O + A_h + A_e$, $O + A_h + A_e + B_h + B_s$, $O + A_h + A_e + B_h + B_s + C$), were treated with model rainwater adapted to local mean rainwater composition. The seepage water was collected. The dissolved organic matter of these soil solutions was fractionated by applying a modified gel filtration technique using Sephadex G 15 (Scholtis 1986). The seepage water was analyzed with a Perkin-Elmer ICP/AES 6000.

3.3 Batch and Columm Experiments

The sorption of specified metal ions from aqueous solutions by humic acids (HA) was studied with the help of batch (static method) and column experiments (dynamic method). The results obtained by these two methods are the maximum sorption capacity (k_1) and the sorption binding constant (k_2), the latter being equivalent to the complex stability constant K [Eq. (2)] of a solid complex-forming substance (Becker 1987).

Both in the batch experiments and in the column experiments, the pH of the solutions was not kept constant, but it was measured at the beginning and the end of each batch experiment, and was measured constantly during the column experiments. The pH decreased from pH 5 to values between 3 and 4 due to the release of protons by the sorption processes.

3.3.1 Batch Experiments

The batch experiments were performed with 30 ml equilibrated solution, 10 mg humic acid, 25°C; contact and shaking time was 24 h.

The sorption parameters k_1 and k_2 were derived from the batch experiments by linear regression from the slope and the axial section of the linearized Langmuir equation. From the regression calculations relative errors of about 2% for the maximum sorption capacity k_1 and about 25% for the sorption binding capacity constants k_2 were deduced (Becker 1987).

3.3.2 Column Experiments

The columns used were glass tubes of 3 mm width filled with humic acids of various particle sizes and conditioned with bi-destilled water. The amount of humic acid is equal to that of the batch experiments. The metal solutions [total concentration c_t (Mn) $= 9.1 \times 10^{-4}$ mol l^{-1} and c_t(Cu) $= 7.9 \times 10^{-4}$ mol l^{-1}] are continuously flowing through the columns; fractioned samples were collected at the outlet and the metal concentrations of the single fractions were measured. The breakthrough curves were used to calculate the maximum sorption capacity as the sum of the retained amounts of metal.

4 Results

4.1 Monolith Lysimeter Experiments

Gel-chromatographic analysis of the dissolved organic substances from lysimeters of different length show a decrease of dissolved substance content with depth. Only organic substances with low molecular volume are transported to lower soil horizons (Fig. 1). This is consistent with earlier observations of Pekdeger (1977). Combined gel-chromatographic, molecular volume separation of dissolved organic substance and the heavy metal analysis in the molecular volume fractions show different affinities of metals either to low or to high molecular volume organic substance (Gruhn et al. 1985; Scholtis 1986). Al, Cr, Cu, Fe and Pb show a strong affinity both to high and low molecular volume fractions (Fig. 2). Ba, Ca, Mg, Mn, Sr and Zn are bound mainly to low molecular volume fractions. Especially the contents of heavy metals which are bound to high molecular volume organic substances decrease therefore rapidly with depth (Fig. 3). These metals, when precipitated within the soil column, are fixed and very immobile (Isenbeck et al. 1985).

The distribution of heavy metals after an intentionally high contamination of soil monolith lysimeters with the heavy metals Tl, Zn and Cu confirms these results. The heavy metals are fixed mainly at the organic rich A_h- and B-horizons of the investigated podzol, which is indicated by higher retardation factors in the A_h- (Zn: 910, Tl: 8110, Cu: 70400) and $B_{h,fe}$-horizons (Zn: 1010–1540; Tl: 1710–1880, Cu: 2230–10000) than in the C-horizon (Zn: 170, Tl: 370, Cu: 220) (Isenbeck et al. 1985).

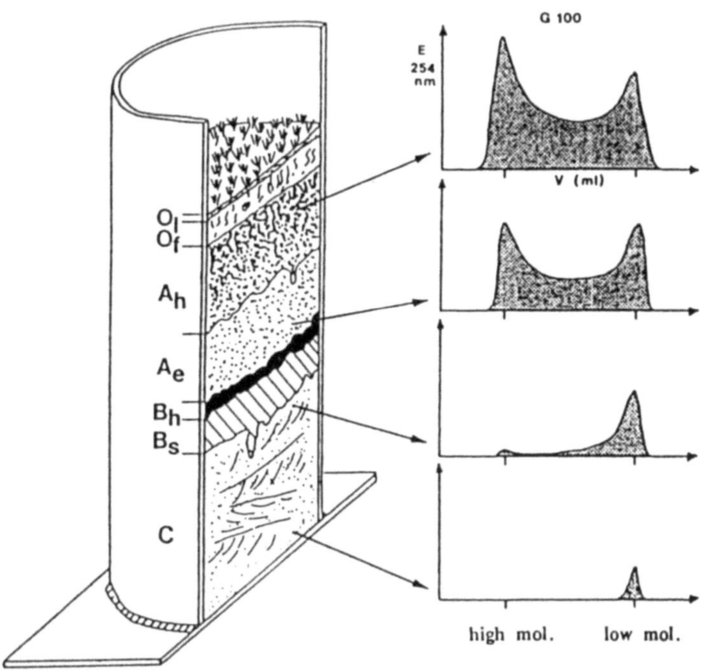

Fig. 1. Molecular volume spectra of dissolved organic substances in seepage water in different soil horizons (fractionation with Sephadex G 100 and deionized water as mobile phase) (Gruhn et al. 1985)

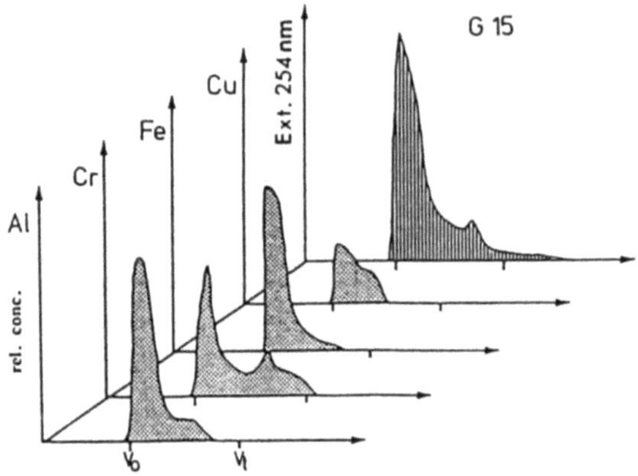

Fig. 2. Gel chromatogram of dissolved organic substances in seepage water from A_h-horizon (Sephadex G15, extinction for light of 254-nm wavelength) and relative metal concentrations in the respective molecular volume fractions (Gruhn et al. 1985)

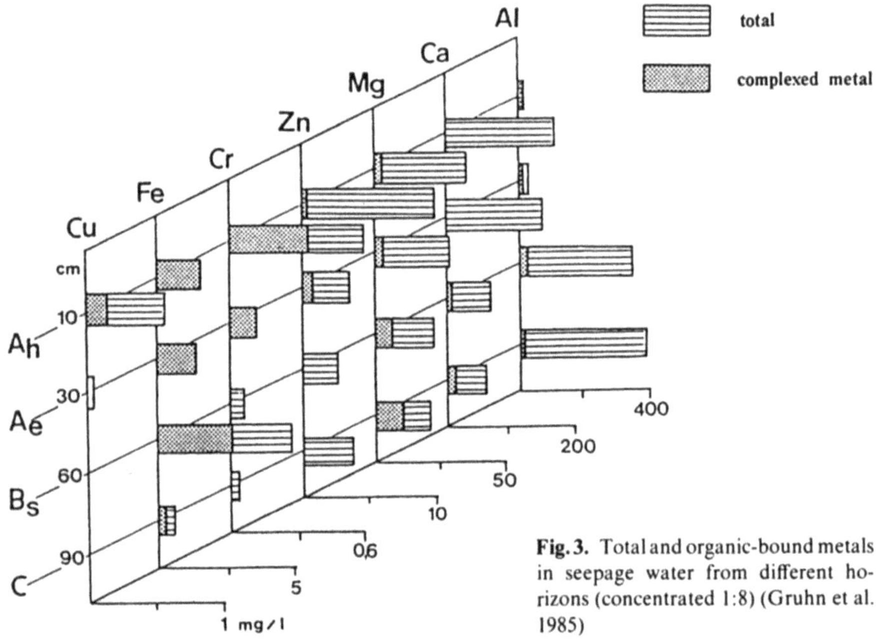

Fig. 3. Total and organic-bound metals in seepage water from different horizons (concentrated 1:8) (Gruhn et al. 1985)

4.2 Batch Experiments

Considering the relative errors, the k_1- and k_2-values obtained by evaluation of linearized Langmuir isotherms (Thielen et al. 1984) within a specified range of copper concentrations do not show significant differences for various particle size fractions of soil humic acids or for those of the two sampled soil horizons (Table 2).

Table 2. Sorption binding constant k_2 and maximum sorption capacity k_1 for copper and humic acids (HA) of different particle size fractions of two soil horizons. Total concentration of copper C_t (Cu): 8.3–333.3 mg l^{-1}

Particle size fraction (μm)		k_2 ($10^4 \times$ l/mol)	k_1 (mol/100 g)
HA Ah (3)			
	32– 80	9.1	0.079
	80–125	8.9	0.080
	125–200	8.9	0.080
HA Bh (3)			
	32– 80	7.9	0.079
	80–125	7.3	0.081
	125–200	8.9	0.080

If the concentrations of the metal ions are varied over a broader range, the results indicate a strong concentration dependency of k_1 and k_2 (Table 3). This shows that for these materials the Langmuir isotherm holds only for certain concentration ranges. Obviously at low metal concentrations only those binding sites are occupied, from which strong binding forces are exerted. With growing metal concentrations also weakly binding sites are occupied, which cause an increase to the maximum sorption capacity and vice versa a decrease of the binding constant. It appears reasonable to describe the concentration dependency of the binding constant of humic substances analogously to the stability function introduced by Gamble et al. (1980) for the description of the formation of metal-humato complexes with the help of sorption functions.

Table 3. Sorption binding constant k_2 and maximum sorption capacity k_1 for different ranges of total metal concentrations C_t (M)

C_t(M) (mg l^{-1})	k_2 ($10^4 \times$ l/mol)	k_1 (mol/100 g)
HA Bh (2); 32–80 μm; calcium		
2.7–16.7	18.1	0.050
2.7–43.3	2.0	0.073
2.7–66.7	0.8	0.092
HA Bh (2); 32–80 μm; copper		
8.3–13.3	19.9	0.059
8.3–16.7	14.2	0.064
8.3–33.3	5.0	0.083
6.7–166.7	1.4	0.101
6.7–333.3	0.6	0.125

4.3 Column Experiments

From the breakthrough curves (Fig. 4) the maximum sorption capacity was calculated as the sum of the retained amounts of metal. The sorption capacities derived from batch and column experiments (Table 4) show that the values obtained by the two methods fit fairly well.

The sorption binding constants were determined with the help of the mathematical simulation of the chromatographic processes in the columns by a black box model (Glueckauf 1955), which has been used by Spies (1984) for the evaluation of the sorption behaviour of sediments. The simulation is performed with assumed data for the sorption binding constant, and for the number N of the theoretical plates of the columns as variables, and with the real concentration of the metal solution and the maximum sorption capacity of the sample of humic acid used in the experiment. After each simulation step the metal ion concentration $[M]_N$ in the N-th segment equals that of the seepage at the end of the column. The optimal adjustment of the simulated to the measured curve is achieved by iterative variation of the sorption binding constant depending on the calculated least square sum of simulated and measured values. In a second program the number of plates is determined by adjusting simulated breakthrough curves to the observed values (Glueckauf 1955). The procedure of a calculation of the simulated breakthrough curves, the parameters used and the resulting sorption binding constants are extensively discussed in Becker (1987).

Figures 5 and 6 indicate the results of the simulation of measured breakthrough curves for the metals manganese and copper. The curves may be simulated

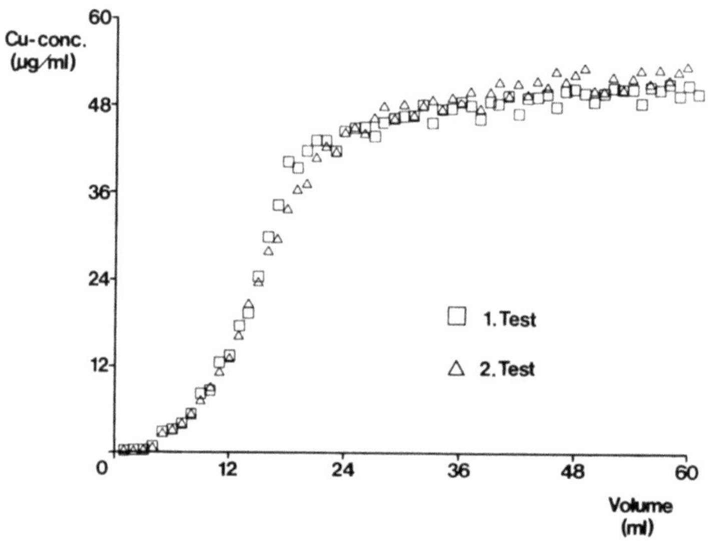

Fig. 4. Breakthrough curves of copper in 3-mm columns filled with humic acids (size 32–80 μm) from the B_h-horizon

Table 4. Comparison of maximum sorption capacities for copper at humic acids (HA) determined with the batch and the column method

Particle size fraction (μm)	Batch method k_1/(mol/100g)	Column method k_1/(mol/100g)
HA Ah		
32– 80	0.079	0.092
80–125	0.080	0.10
125–200	0.080	0.093
HA Bh		
32– 80	0.079	0.090
80–125	0.081	0.106
125–200	0.080	0.084

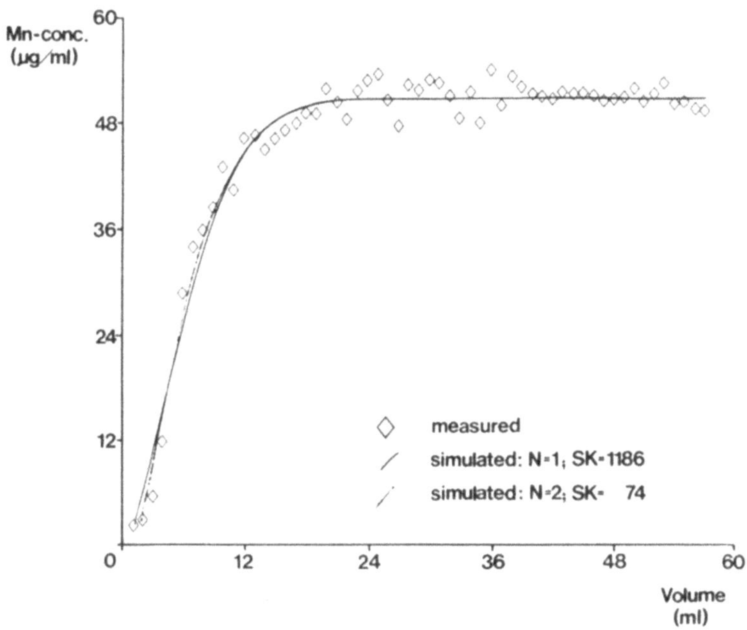

Fig. 5. Simulated and measured breakthrough curves of manganese in 3-mm columns filled with humic acids (size 32-80 μm) from the B_h-horizon [sorption capacity 0.063 mol/100 g; c_t (Mn) = 9.1 × 10^{-4} mol l^{-1}]

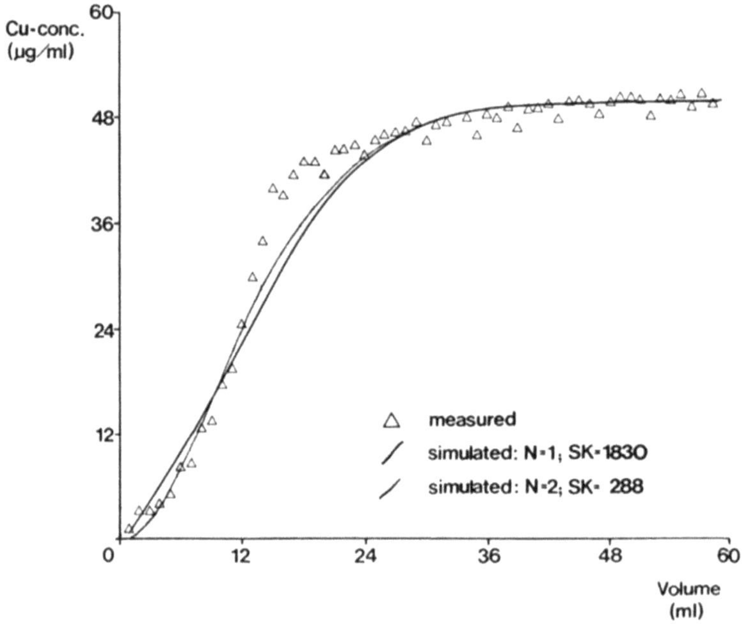

Fig. 6. Simulated and measured breakthrough curves of copper in 3-mm columns filled with humic acids (size 32–80 μm) from the B_h-horizon [sorption capacity 0.108 mol/100 g; c_t (Cu) = 7.9×10^{-4} mol l^{-1}]

either with a high sorption binding constant and one plate, or with a lower sorption binding constant and two plates. If simulated with one plate, sorption binding constants of 1186 $l \times mol^{-1}$ for manganese and 1830 $l \times mol^{-1}$ for copper are obtained. The simulation with two plates indicates for copper 74 $l \times mol^{-1}$ and for manganese 288 $l \times mol^{-1}$. At the present state of knowledge it cannot be explained how the sorption binding constant and the number of plates interact and in which respect the resulting values are influenced. The mathematical simulation of the manganese curve (Fig. 5) fits better to the observed values than that for the copper curve (Fig. 6). This deviation may be qualitatively explained by assuming different, strong binding forces at specific sites of the humic acid molecules. Manganese is fixed mainly on the strong binding sites of the humic acids, whereas copper forming more stable complexes is fixed as well in strong as in weak binding sites. The first part of the breakthrough curve (Fig. 6) describes the reactions of the stronger binding sites, whereas the last part of the curve depicts also the reaction of the weakly complexing sites. Following the procedures mentioned above, a better fit between the measured and calculated curve (Fig. 7) is achieved for the initial part of the breakthrough curve for copper by a simulation with a 10% lower sorption capacity than the maximum sorption capacity. The sorption binding constant obtained from the simulated curve, representing the strong binding sites, is in this case about three times higher (k_2 = 5282) than the constant obtained by fitting the whole curve (k_2 = 1830).

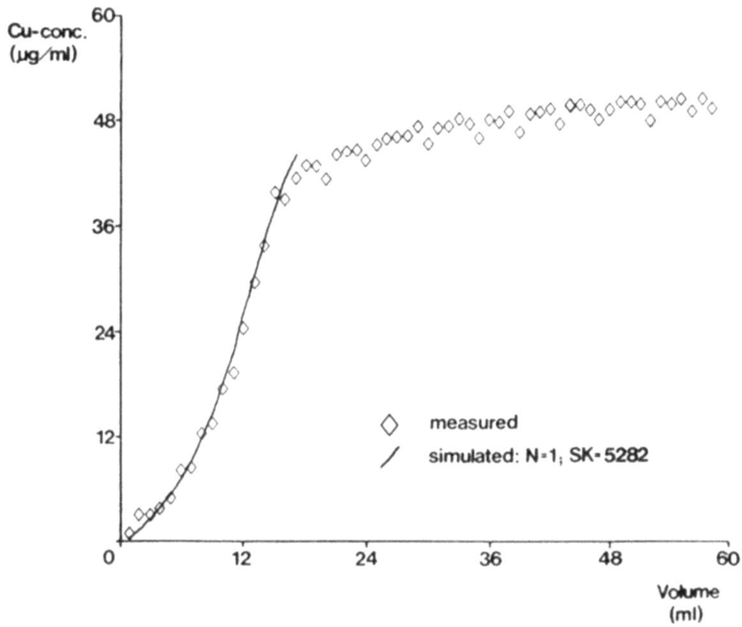

Fig. 7. Simulated and measured breakthrough curves of copper in 3-mm columns filled with humic acids (size 32–80 µm) from the B_h-horizon, using a 10% lower sorption capacity than the maximum sorption capacity in Fig. 6 [c_t (Cu) = 7.9×10^{-4} mol l^{-1}]

The results of the column experiments are in good agreement with the results obtained with batch experiments. There is again a concentration dependence of the sorption properties of humic acid.

5 Conclusions

The transport of toxic metals in the unsaturated and saturated zones is strongly affected by dissolved and solid organic substances, which alter their sorption behaviour and their solubility. The adsoption of metals on solid organic substances retards their movement in comparison to seepage and groundwater. Irreversible, specific adsorption to solid organic substances removes relatively high percentages of dissolved metal species from the pore solutions. The solubilities of the metals are furthermore increased or decreased in the presence of complex-forming organic substances. Compared with the importance of this subject very few quantitative data are available.

The interaction between dissolved and solid organic substances and various metals (Al, Cd, Co, Cr, Cu, Fe, Mn, Ni, Pb and Zn) has been investigated with respect to their movement or fixation in the different horizons of a podzol profile on Quaternary fluvio-glacial sediments. The comparison between the metal contents bound to dissolved and to solid organic substances indicate the organo-metal complexes as mobility-controlling factors of inorganic toxic elements in such porous media. Detailed studies of the nature of the respective organic substances and of the stability constants of these complexes aim to explain the relative importance of the different types of interactions. Future work is needed to implement the obtained data into existing geochemical and transport models and to compare the model results with the field observations. Considering the importance of the organic substances for the understanding of the transport behaviour of toxic metals in subsurface water, more laboratory and field research is necessary.

References

Becker R (1987) Untersuchung zum Sorptionsverhalten von Boden-Huminsäuren gegenüber Metall-Ionen mit der statischen und dynamischen Methode. Ph D Thesis, Univ Marburg

Chen Y, Senesi N, Schnitzer M (1977) Information provided on humic substances by E_4/E_6-ratios. Soil Sci Soc Am J 41:352–358

Flaig W, Beutelspeicher H, Rietz E (1975) Chemical composition and physical properties of humic substances. In: Gieseking IE (ed) Soil components I. Organic components. Springer, Berlin Heidelberg New York, pp 1–211

Förstner U, Müller G (1974) Schwermetalle in Flüssen und Seen. Springer, Berlin Heidelberg New York, 225 pp

Gamble DS, Underdown AW, Langford CH (1980) Copper (II) titration of fulvic acid ligand sites with theoretical, potentiometric, and spectrophotometric analysis. Anal Chem 52:1901–1908

Garrels RM, Christ ChL (1965) Solutions, minerals, and equilibria. Harper & Row, New York Evanston London; Weatherhill, Tokyo, 450 pp

Glueckauf E (1955) Theory of chromatography. Part 9. The "theoretical plate" concept in column separations. Trans Farad Soc 51:34–44

Gruhn A, Matthess G, Pekdeger A, Scholtis A (1985) Die Rolle der gelösten organischen Substanz beim Transport von Schwermetallen in der ungesättigten Bodenzone. Z Dtsch Geol Ges 136:417–427

Isenbeck M, Schröter J, Kretschmer W, Matthess G, Pekdeger A, Schulz HD (1985) Die Problematik des Retardationskonzeptes – dargestellt am Beispiel ausgewählter Schwermetalle. Meyniana 37:47–64

Jackson RE (ed) (1980) Aquifer contamination and protection. Studies A, UNESCO, Paris. Hydrol 30:439 pp

Jackson RE, Merritt WF, Champ DR, Gulens J, Inch KJ (1980) The distribution coefficient as a geochemical measure of the mobility of contaminants in a groundwater flow system. IAEA, Vienna. Panel Proc Ser STI/PUB/518:209–225

Matthess G (1984a) Unsaturated zone pollution by heavy metals. In: Yaron B, Dagan G, Goldshmid J (eds) Pollutants in porous media. Springer, Berlin Heidelberg New York Tokyo, pp 79–93

Matthess G (1984b) The role of natural organics on water interaction with soil and rock. Int Assoc Hydrol Sci 150:11–21

Matthess G, Pekdeger G (1980) Chemisch-biochemische Umsetzungen bei der Grundwasserneubildung. Gas Wasserfach 121:214–219

Matthess G, Pekdeger A, Schulz HD (1977) Geochemical-biogeochemical processes in seepage water during groundwater recharge. Proc Symp Water-rock-interaction, Strasbourg 1, pp 146–155

Matthess G, Isenbeck M, Pekdeger A, Schenk D, Schroeter J (1985) Der Stofftransport im Grundwasser und die Wasserschutzgebietsrichtlinie W 101. Umweltbundesamt Berichte 7. Schmidt, Berlin, 181 pp

Parkhurst DL, Thorstenson DC, Plummer LN (1980) PHREEQE – a computer program for geochemical calculations. US Geol Surv Water Res Inv 80-96:210 pp
Pekdeger A (1977) Labor- und Felduntersuchungen zur Genese der Sicker- und Grundwasserbeschaffenheit. Ph D Thesis, Univ Kiel
Plummer LN, Jones BF, Truesdell AH (1976) WATEQF – a Fortran IV version of WATEQ, a computer program for calculating chemical equilibrium of natural waters. US Geol Surv Water Res Inv 76/13: 615 pp
Post B (1987) Chromatographische und spektroskopische Charakterisierung von Boden-Huminstoffen nach sukzessiver Lösungsmittel-Extraktion. Ph D Thesis, Univ Marburg
Scheffer F, Schachtschabel P (1982) Lehrbuch der Bodenkunde, 8th edn. Enke, Stuttgart, 442 pp
Schnitzer M, Khan SU (eds) (1978) Soil organic matter. Developments in soil science 8. Elsevier, Amsterdam, 319 pp
Scholtis A (1986) Transport von anorganischen und organischen Substanzen durch die wasserungesättigte Zone unter der Einwirkung unterschiedlicher Modell-Niederschlagswässer. Ph D Thesis, Univ Kiel
Schulz HD (1970) Chemische Vorgänge beim Übergang vom Sickerwasser zum Grundwasser. Geol Mitt 10:151-204
Spies RH (1984) Charakterisierung der Sorptionseigenschaften eines Sediments des Roten Wassers durch Trennstufenzahl, Sättigungskapazität und Sorptionskonstante. Ph D Thesis, Univ Marburg
Stumm W, Morgan JJ (1981) Aquatic chemistry, 2nd edn. John Wiley & Sons, New York, 780 pp
Thielen H, Bodes H, Lorenz I, Klamberg H (1984) Wechselwirkungen zwischen Huminstoffen und Metall-Ionen in wässrigen Systemen. Proc Int RIZA Symp, Munich, pp 579-589
Wedepohl KH (ed) (1978) Handbook of geochemistry. II-1. Springer, Berlin Heidelberg New York

Suitability of Sorption-Desorption Models to Simulate Partitioning and Movement of Ions in Soils

N.J. Barrow[1]

1 Introduction

If we are to describe the movement of ions in soil, we need to understand the partitioning of the ions between the moving and the stationary phases, i.e. between the solid and the liquid phases. Furthermore, if we either increase or decrease the concentration in the liquid phase, we need to know to what extent this change will be modified by sorption or desorption.

There was a time when it all seemed simple. We thought that ions were *ad*sorbed, equilibrium was rapidly reached, and desorption was therefore a simple reversal of the original adsorption. These simple ideas should have been abandoned when it was found that desorption did not, in fact, follow the same track as "ad"-sorption. In this chapter I shall use "sorption" to mean any process by which ions are removed from solution. The word is meant to be descriptive and to imply no mechanism. Note that sorption includes adsorption.

We should also have abandoned the notion that equilibrium was rapidly achieved. There is abundant evidence (see later) that reaction continues, albeit slowly, for a long time. This too should have shown that the process is not a simple adsorption. If it were it should have been completed fairly quickly, at the most, a matter of minutes when soil and solution were adequately mixed. On the other hand, there is much evidence that the initial reaction is between ions in solution and the surface of soil particles, i.e. an adsorption reaction. Hence, some other reaction or process must follow the initial adsorption. If we are to describe sorption-desorption our description must include a component to account for these effects. This means that we must be able to account for the effects of time. Four kinds of approach to this problem may be discerned: simple, constrained, pragmatic, and mechanistic. These are considered next.

2 Simple Approaches

The simplest approach is to ignore the argument in the previous section and to assume, either explicitly or implicitly, that the reaction can be represented as:

[1]CSIRO Division of Animal Production, Private Bay, PO Wembley 6014, Western Australia, Australia

$S+I = SI$ (where S represents a surface site and I represents an ion). This may then be simplified further by assuming that $S \gg I$, and that the back reaction can be ignored. Thus, the assumption is that the rate is proportional to the concentration of I, a first-order reaction. This simple approach was used, for example by Aringhieri and Pardini (1983) to describe the reaction of OH^- ions with soil. As might be expected, it only described the results over a brief period. A more general treatment is to regard the reaction as a second-order forward reaction opposed by a first-order back reaction. At equilibrium such a reaction gives rise to the Langmuir adsorption equation and the kinetics are often referred to as Langmuir kinetics. Such equations were applied by Aringhieri et al. (1985) to the reaction of copper and of cadmium with an Italian soil. They did not describe the rate, and indeed the rate appeared to decrease with time. A somewhat expanded version of this approach was used by Amacher et al. (1986) who included some further variations of single-reaction models. None of them were adequate. There are two reasons why these simple approaches do not work. The main one is that, for periods of most practical interest, the rate-limiting step is not the initial reaction but the slow process that follows it. A less important reason is that the reaction is indeed between ions and charged surfaces. The charges on the ions and the charges on the surfaces affect the rates. This aspect, though of considerable scientific interest, is of little practical concern and will not be considered further.

The Elovich equation may also be considered under this heading. This equation has been advocated by Chien and Clayton (1980) and by Hingston (1981). It was originally considered to be empirical but Atkinson et al. (1971) showed that it could be derived by assuming a heterogeneous distribution of activation energies of the adsorbing sites. Thus, the basic postulate appears to be that the reacted material remains on the surface. To test this equation, sorption is plotted against log time. This plot only describes the results under very restricted circumstances (Barrow 1983a). Worse, from the point of view of modelling the movement of sorbates in soil, it only purports to describe the changes in concentration for one level of addition. That is, it relates concentration to time, whereas we need a function that inter-relates the three variables: concentration, time, and the amount sorbed.

3 Constrained Approaches

In these approaches it is recognized that the initial adsorption reaction is followed by further reactions, however, the kinds of further reactions that are considered are those that can be incorporated into the continuity equations that are used to describe the movement of water and of solutes through soil columns. For example, Mansell et al. (1985) used a model in which first-order reactions described the transfers between "chemically adsorbed" and "physically adsorbed" phosphate. They tested this model at one level of addition of phosphate for up to 36 h. The model gave a poor description of leaching of phosphate through soil. Novak and Petschauer (1979) used a model in which an initial adsorption reaction was followed by an irreversible surface reaction. They found that they could describe the reaction with phosphate for up to 24 h. However, the values of the coefficients they used were such that after a few days almost all of the added phosphate would be in the irreversibly bound form. If this were really the case there would be little point in applying phosphate fertilizers! Thus, an extrapolation of these results to realistic periods shows that the model is unsuitable. A similar approach was used by Lin et al. (1983) who postulated that an initial adsorption reaction was followed by a reversible reaction in which "labile" phosphate is converted into "non-labile" phosphate. The terminology used is, to say the least, puzzling. If the reaction is reversible how can the phosphate be called non-labile? However, the main problem with these approaches is that they do not describe the observations very well; the continuing reaction simply does not fit first-order kinetics. One possibility for overcoming this problem is to add a sequence of further reactions. In a somewhat different context, Probert and Larsen (1972) discussed the merits and demerits of this kind of approach. They pointed out the dangers when the implied products cannot be identified and that the longer the experiment runs, the more terms that must be invoked. They therefore preferred a simple two-constant equation similar to that introduced in a later section.

A chain of first-order reactions has some similarities to a diffusion process and it has often been suggested that a diffusion step was involved in the continuing reaction with soil. Sometimes this has been merely suggested and not tested. For example, Ibrahim and Pratt (1982) suggested that the slow reaction involved "penetration and adsorption of P into porous solids". Sometimes it has been tested, but inappropriately. Under certain conditions, the amount of a substance diffusing is proportional to the square root of time. One of the conditions is that diffusion should be into a plane. For soil particles this is clearly not so. However, if the distance of diffusion is small relative to the particle size, this is an acceptable approximation. Another restriction is that the source concentration should be constant. If it is postulated that the rate-limiting step is from the solution to the solid phase, then the concentration in the solution phase should be constant. This will not be so if soil is simply mixed with a solution containing the reacting material, as was assumed by Tambe and Savant (1978). If, however, it is postulated that the rate-limiting step is from the surface of the solid phase into its interior, then it is not sufficient that the solution concentration be constant. This is because we are considering the reactions of ions with charged surfaces. If some of those ions penetrate the surface, its electrical potential will change. Consequently, the amount of *ad*sorbed ions will change. Hence, the source concentration for the diffusion will not be constant even if the concentration in solution is constant. Further consideration will be given to diffusion under the heading of mechanistic approaches.

4 Pragmatic Approaches

In a strictly pragmatic approach no attention is given to possible mechanisms and the aim is merely to describe the observations. This may be justified by the argument that, given an accurate description of the processes, we would be in a better position to deduce the mechanisms. An approach that has proved effective involves two assumptions. One is that adsorption is rapid and that the resulting equilibrium between the concentration in solution and the *ad*sorbed form can be described by the Freundlich equation. The other assumption is that the amount that remains in the *ad*sorbed form decreases with time (t) according to st^{-b2} (where s is the amount in the sorbed form and therefore includes the adsorbed form). Thus:

$$a\,c^{b1} = s\,t^{-b2} \tag{1}$$

and

$$s = a\,c^{b1}\,t^{b2}. \tag{2}$$

This equation was introduced by Kuo and Lotse (1974) and was also used by Barrow and Shaw (1975a). In both cases the model was derived from assumptions that were intended to imply a mechanistic base, but it is probably better to regard it as a simple, pragmatic equation. Unfortunately, Kuo and Lotse (1974) modified Eq. (2) to:

$$s = a\,c_0^{b1}\,t^{b2}, \tag{3}$$

where c_0 is the initial solution concentration. This modification permits simple linear regressions to be used for each initial level of application. In this form, it has been used to describe sorption of phosphate by Schwertmann and Schieck (1980), of arsenite by Elkhatib et al. (1984), and of sulphate by Singh (1984). However, the supposed simplification leads to unnecessary complexity, such as the calculation of rate coefficients at different times (Singh 1984) and the calculation of different coefficients for different initial concentrations (Schwertmann and Schieck 1980; Elkhatib et al. 1984).

In contrast, Eq. (2) has only three coefficients and these can be easily estimated using multiple regression on the log form. It closely described phosphate sorption on a range of Australian soils (Barrow and Shaw 1975a) and on a range of world soils (Barrow 1980a,b). It also describes molybdate, fluoride, and sulphate sorption (Barrow and Shaw 1975b, 1977a,b). Such wide applicability shows that this simple equation is often a very effective summary of behaviour. One would therefore expect that it would be widely used, for example in modelling the movement of materials through soil. Three reasons may be suggested why this is not so. Firstly, the equation does not adequately describe the behaviour of all reactants. In Eq. (2), it is assumed that the Freundlich equation describes the relation between the adsorbed form and the concentration in solution. This means that log-log plots of sorption against concentration are linear. In many cases this is an acceptable approximation over important concentration ranges. But if the range of concentrations is much bigger, log-log plots may be curved. This is often especially noticeable for the reaction of metals such as zinc with soil. For example, marked curvature of log-log plots was observed by Elrashidi and O'Connor (1982) who attempted to describe it by dividing the curve into two linear segments. There may be several reasons for these shapes. One is that there is no special reason why a Freundlich equation should be precisely followed. A second is that there may be some contribution from sorbate already present and this changes the shape of log-log plots (Barrow 1978). And a third is that reaction with metals decreases the pH and this induces a feedback effect decreasing further sorption. Sorption can usually be *described* by replacing the Freundlich relation on the left-hand side of Eq. (1) by a more appropriate relation, such as the equations suggested by Sibbesen (1981), and, where appropriate, adding a term for the sorbate already present (q). This would give:

$$a\, c^{b_1 c^{-b_3}} = s\, t^{-b_2} + q. \tag{4}$$

A second reason can be understood if Eq. (2) is differentiated with respect to time:

$$\delta s/\delta t = a\, b_2 c^{b_1}\, t^{b_2 - 1}. \tag{5}$$

Because b_2 is less than unity this equation tells us that the rate of the reaction decreases with time. This means that for a given addition of a reactant, the current rate of change depends on the period that has elapsed since that addition was made. This is an important and informative summary of the behaviour and it explains why simple approaches involving first-order reaction schemes do not work: the effect of time is different from that which they describe. It also shows why such descriptions have not been incorporated into mechanistic models of movement of reactants. These models seem to require simpler equations in which the rate of change is not dependent on time.

A third reason why this approach has not been widely adopted is that Eq. (2) describes the changes in sorption after one addition of reactant. Most practical problems are more complex. They may involve repeated additions or they may involve removal of a previous addition, i.e. desorption. It is not widely realized that these aspects are really the opposite sides of the same coin. Suppose we have added a reactant to a soil and have let it react for a long period. Then we may choose to decrease the concentration in the soil solution and thereby induce desorption; or we may choose to increase the concentration and thereby induce further sorption. The effects of these two possible operations are symmetrical (Barrow 1983b). That is, plots of desorption against concentration differ from plots of the original sorption but they are consistent with plots of new sorption against concentration. The question which then arises is how to modify the simple pragmatic equations so that they may describe these more complex effects. However, before considering this, it is desirable to consider mechanistic descriptions of the reaction of ions with soil.

5 Mechanistic Approaches

To be acceptable, a mechanistic approach should be explanatory, effective, comprehensive, and soundly based. Explanatory means that it should lead to further understanding of the process. Effective means that it closely describes the observations; this is an essential, though not a sufficient, criterion for all mechanistic approaches. Comprehensive means that it describes all of the observations. And soundly based means that it is consistent with detailed knowledge of the processes that are involved. Of the four criteria, that of comprehensiveness has been the most difficult to achieve. Many of the supposed explanations that have been offered to describe observations in soil chemistry are appropriate only to limited conditions, for example, to the effects of pH on one reactant. To be comprehensive, an explanation must account for all of the observations. These have been discussed in detail by Barrow (1987) and only a brief summary is given here. Ten separate effects can be identified when dilute solutions containing a single reactant react with soils.

1. There is a characteristic relation between the amount sorbed and the concentration remaining in solution.
2. Reaction continues for a long time at a rate that is proportional to a small fractional power of time.
3. If this slow reaction is thought of as slow conversion to a firmly held form, then the amount so converted is proportional to the amount added.
4. Increasing the temperature increases the rate of sorption; increasing the temperature therefore decreases the concentration in solution.
5. Increasing the temperature increases the rate of desorption.
6. When the rate of change is very slow, increasing the temperature changes the equilibrium between solution and soil so that the concentration in solution increases, i.e. the *ad*sorption reaction is exothermic.
7. Repeated additions behave as separate pulses; they follow the same rules as single additions except that the instantaneous buffering capacity is decreased.
8. Desorption can be treated as a negative addition and the effects of desorption and sorption of further additions are symmetrical.
9. There are characteristic effects of pH for each reactant.
10. The effects of pH interact with the nature and concentration of the background electrolyte.

In order to describe these diverse observations, three assumptions are needed. They are: that *ions* react with *variable-charge* surfaces; that the surfaces are heterogeneous; and that adsorption is followed by diffusive penetration of the surfaces. These assumptions can be precisely expressed as mathematical equations and the resulting system of equations combined into a model. The behaviour of that model, the way that it explains the ten observations on sorption, and the way that it meets the four criteria for a mechanistic model, have been dealt with in detail by Barrow (1987). Only three aspects will be mentioned here.

The first aspect is the effects of pH. For all reactants, the effects of pH on sorption are described by the interaction between the effects of pH on the ions present in solution and on the electric potential of the surface. The model explains in a consistent way the diverse effects of pH on anions such as phosphate, fluoride, borate, molybdate, selenite, selenate, and sulphate and metals such as zinc. For example, for phosphate, the model postulates that the reaction with the surface is dominated by the divalent ion HPO_4^{2-}. On the one hand, the concentration of this ion in solution increases with increasing pH but, on the other hand, increasing pH makes the surface potential more negative. This has a large effect on the divalent ion and the net effect is usually a decrease in sorption with increasing pH. Much the same explanation applies to selenite. Again it is postulated that the divalent ion (SeO_3^{2-}) dominates the reaction. The pK_2 for selenious acid is about 8.5 and so the concentration of (SeO_3^{2-}) also increases with increasing pH. The net effect is rather like that for phosphate, i.e. a decrease in sorption with increasing pH. For borate, the pK is at about 9.2 and so, again, the concentration of the relevant ion, in this case $B(OH)_4^-$, increases with increasing pH. However, for borate, the ion is monovalent and the effects of the electrical potential are therefore smaller. When the pH is increased, the adverse electrostatic effects on the surface are not large enough to oppose the effects of the increasing concentration of borate ions. Sorption of borate therefore increases. For fluoride, selenate, and sulphate, the relevant pKs are low and the ions are fully dissociated at soil pH values. With increasing pH, the increasingly negative potential of the reacting surfaces is not opposed by an increasing concentration of the reacting ion and sorption generally decreases. Fluoride is somewhat of an exception because fluoride ions form strong complexes with aluminium ions in solution and this tends to decrease fluoride sorption at low pH. For metals such as zinc, the sheath of water molecules around the divalent ions is disturbed when one of the water molecules dissociates to form, in effect, $MeOH^+$ ions. It is postulated that this disturbance makes these ions much more likely to react. Their concentration increases with increasing pH and this is a major factor in increasing the sorption with increasing pH.

The electric potential is also affected by the nature and concentration of the background electrolyte. The model includes these effects to explain why increasing the salt concentration can increase anion sorption at moderate pH but *decrease* it at low pH. Increases in the salt concentration increase the concentration of charge-balancing ions near the surface and this results in a decrease in the absolute value of the electric potential. If the potential is positive, as at low pH, it becomes less positive and adsorption of anions is decreased. If the potential is negative, as at higher pH, it becomes less negative and adsorption of anions increases.

The second aspect considered is the effects of time. The model postulates that the continuing reaction is due to slow, diffusive penetration of the adsorbing surface. Provided that the source concentration is constant and that the surface can be approximated to a plane, the rate of diffusion is proportional to the square root of time. However, the source concentration for diffusion is the adsorbed reactant. This concentration is not constant even when the solution concentration of the reactant is constant. This is because ion penetration changes the electric potential of the surface and this, in turn, changes the balance between adsorbed and solution ions. This "drag" effect decreases the exponent for time below that appropriate to a square root (0.5). Thus, the model is consistent with the observation that the rate of change is proportional to a small fractional power of time.

The third aspect considered is the effect of repeated applications. An important component of the diffusion model is that each change in the source concentration causes a pulse or a ripple which is then propagated down the diffusion gradient. The formal description is that there is a linear superposition of a sequence of diffusion equations. The practical outcome is that separate additions of reactant are treated separately but the effect of each addition differs from that of the previous one because the continuing diffusive penetration has changed the electric potential of the surface.

An acceptable mechanistic model is of great value for understanding the processes that are involved. However, the computations involved in such models are complex and they are not readily adapted for use in calculating movement of reactant in practical cases. For this purpose, simpler models have advantages. Nevertheless the simpler models should be compatible with the mechanistic models. Thus, for example, the observation that the rate of reaction is often proportional to a small, fractional power of time should be seen as a convenient abstract of the mechanistic model in which the rate of reaction is determined by diffusion modified by a feedback effect.

6 Simulation of Movement

It is natural that the scientists with most interest in the movement of reactants through soils and aquifers have been highly skilled in the mathematics of hydrology. They have therefore attempted to describe the movement of reacting chemicals using mathematics that were compatible with the mathematics for movement of water. This has led to the constrained approaches described above. However, the pragmatic approaches have shown that the rates of reaction do not correspond to the rates assumed in such approaches, and the mechanistic approaches have shown why this is so. This seems to lead to an impasse in which sophisticated description of water movement is incompatible with an accurate description of the reaction component. One way around this impasse is to revert to layer models in which a series of time steps and a series of depth steps are used to approximate the continuous effects of time and depth. These simple, less mechanistic but functional models are not without advantages (Addiscott and Wagenet 1985). For each layer, there is a series of additions (or removals) of reactant. We therefore need to consider the description of repeated additions of reactant.

It was suggested by Barrow and Shaw (1975a) that each addition of a nutrient should be treated as a separate pulse. As is indicated above, this is consistent with a mechanistic model. Thus, for three separate additions, Eq. (1) becomes:

$$a\, c^{b1} = s_0\, t^{-b2} + s_1(t - t_1)^{-b2} + s_2(t - t_2)^{-b2}, \tag{6}$$

where an amount s_0 is sorbed at $t = 0$ and extra amounts s_1 and s_2 are added at t_1 and t_2, and where t is the total elapsed time. This equation was found to give a fairly good description of the changes in concentration which occurred when repeated additions of phosphate were made (Barrow and Shaw 1975a). However, there were important deviations from the relation predicted by Eq. (6). These occur because reaction inevitably changes the soil; reaction with anions, for example, always makes the electric potential of the reacting surfaces more negative. This means that subsequent additions are made to a different substrate from the original one, i.e. a substrate that has less ability to react than the original one. As a consequence, the observed solution concentrations were higher than predicted by Eq. (6) (Fig. 1). In other words, term a of Eq. (6) is not a constant. Its

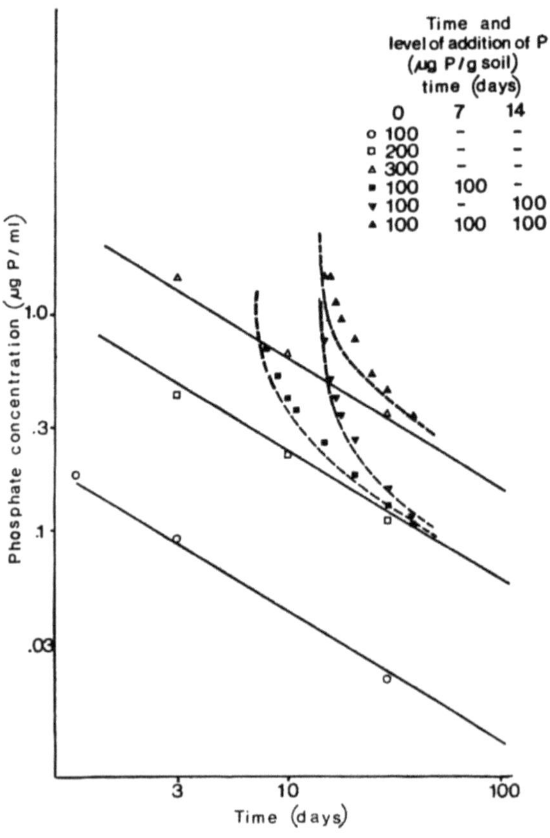

Fig. 1. Effect of repeated applications of phosphate on the subsequent changes in concentration of phosphate in solution. The *solid lines* indicate Eq. (2) which was fitted to values for a single addition. The values of the parameters thus obtained were then applied to Eq. (6) to predict the concentration when repeated additions were made (Barrow and Shaw 1975a)

value depends on how much of the reactant had previously been added to the soil (Fig. 2). It also depends on the period that has elapsed since the reactant was added (Fig. 3). To illustrate the effects of this, let us consider only one subsequent addition and introduce a term to modify the value of term a when the reaction to this addition is being considered:

$$c^{b1} = \frac{s_0}{a} t^{-b2} + \frac{s_1}{a\,k_1}(t - t_1)^{-b2}, \tag{7}$$

where $k_1 < 1$. The k_1 term has the effect of increasing the concentration. Equation (7) may be rearranged to show the changes in sorption to be expected after the second addition of reactant:

$$s_1 = k_1\,a\,c^{b1}(t - t_1)^{b2} - s_0\,k_1\left(\frac{t - t_1}{t}\right)^b \tag{8}$$

The first term of this equation produces an effect of time which is similar to that on the original addition [Eq. (2)], with the difference that the k_1 term decreases the sorption capacity of the soil. The second term tells us that sorption only occurs if the concentration is above a certain threshold level. This threshold continues to decrease with time because of the effect on the continuing reaction of the original addition. Below this threshold desorption occurs. If we wish to

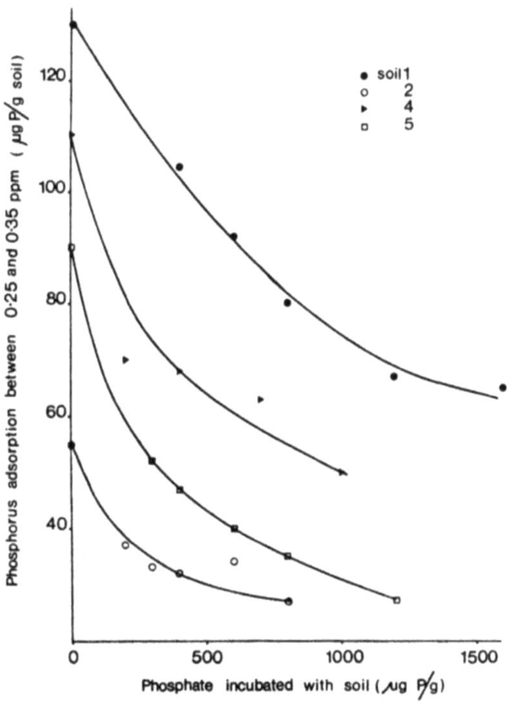

Fig. 2. Effect of incubating soils with the indicated levels of phosphate for a year on the slope of graphs of plots for subsequent sorption of phosphate. These slopes are proportional to the value of $k_1 a$ of Eq. (8) (Barrow 1974)

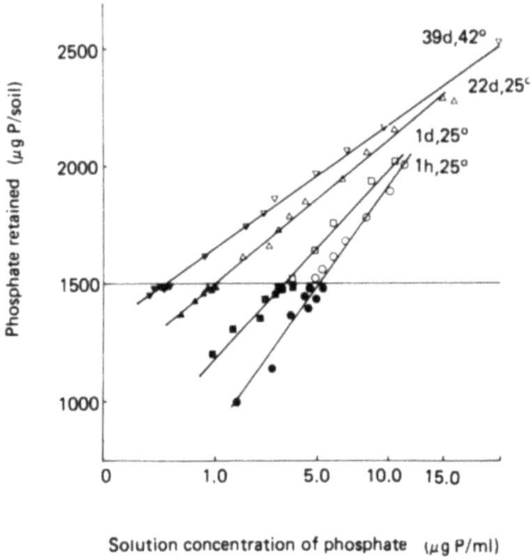

Fig. 3. Effect of incubating phosphate with a soil for the indicated periods on the subsequent sorption (*open symbols*) and desorption (*closed symbols*) of phosphate. Data are plotted against concentration raised to the power b_1. This linearizes Eq. (8) and the slopes reflect values of k_1 a (Barrow 1983b)

specify desorption (s_d) rather than sorption, then the equation is simply the opposite of Eq. (8):

$$s_d = s_0 \, k_1 \left(\frac{t - t_1}{t}\right)^{b2} - k_1 \, a \, c^{b1} \, (t - t_1)^{b2}. \tag{9}$$

The first term of this equation can now be seen to represent the amount of desorption that could occur if the solution concentration were decreased to zero. Inspection of this term shows that appreciable desorption could only occur when the period of desorption (t-t_1) is large relative to the total time (t). Even then, the k_1 term will have an effect and, as its value is less than one, it will decrease desorption.

Equations (8) and (9) are suggested as general, pragmatic equations to describe the behaviour when repeated additions of a reactant are made or when desorption is induced. It is implied that the conditions during the several phases of the experiment are constant. This has not always been the case in experiments in which desorption was studied. Thus, in studies on the desorption of phosphate and of fluoride (Barrow 1979; Barrow and Shaw 1975c, 1977a), the sorption step involved incubating the reactants with the soil but the desorption step involved mixing solution and soil at large solution:soil ratios. Even so, the equations which were found to describe desorption were similar to Eq. (9).

In order to use Eq. (8) in a practical layer model it is necessary to generalize it. Let A_i represent an amount of reactant that enters a layer at time t_i and let s_i be the amount of sorption that has occurred at the subsequent time t. If we consider the second step as in Eq. (8), then the total sorption s_T is given by:

$$s_T = s_0 + s_1, \tag{10}$$

and therefore:

$$s_T = a\,k_1\,c^{b1}\,(t - t_1)^{b2} + s_0 \left(1 - k_1 \frac{t - t_1}{t}\right)^{b2} . \tag{11}$$

Generalizing:

$$s_T = a\,k_i\,c^{b1}\,(t - t_i)^{b2} + \sum_{0,j}^{i} \left[s_i \left(1 - \frac{k_j}{k_i} \frac{t - t_i}{t - t_j} \right)^{b2} \right] . \tag{12}$$

This equation is simultaneous with an equation to partition the reactant between the sorbed and the solution phases:

$$s_T = \sum^{i} A_i - c\,R, \tag{13}$$

where R is the solution:soil ratio. These two equations are solved by finding the value of c for which the two calculated values of s_T agree. An appropriate code[2] has been incorporated into a layer model in order to illustrate the effects of the components of the description of sorption and desorption on the leaching process as indicated by the breakthrough curves (Fig. 4). The experiment simulated is one in which a constant inflow concentration is maintained for 12 pore volumes and this is then replaced by a zero inflow concentration. In each case, the sorption characteristics are chosen so that the efflux concentration approaches the inflow concentration after about 12 pore volumes. The figure shows that if one makes the unrealistic assumption that there is an instantaneous reaction and a linear relation between sorption and concentration, and therefore a constant partition function, the outcome is an almost symmetrical breakthrough curve. If the relation between

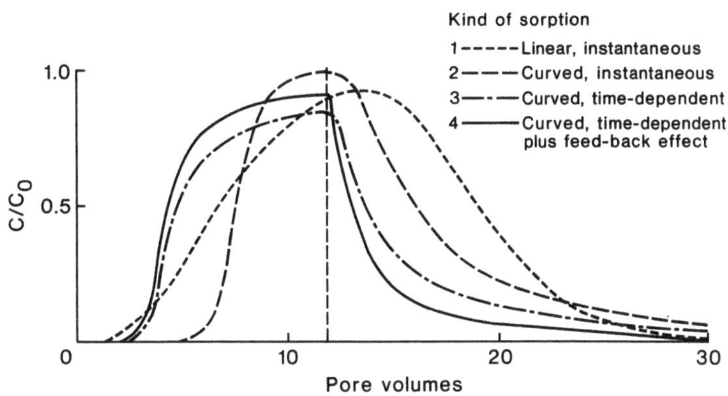

Fig. 4. Simulated breakthrough curves when a solution of initial concentration c_0 enters a wetted soil column and, after 12 pore volumes have been eluted, is replaced by a solution of zero concentration. Curves *1* to *4* represent a sequence of increasing realism in the relation between sorption and solution concentration

[2]The Basic program is available upon request from the author.

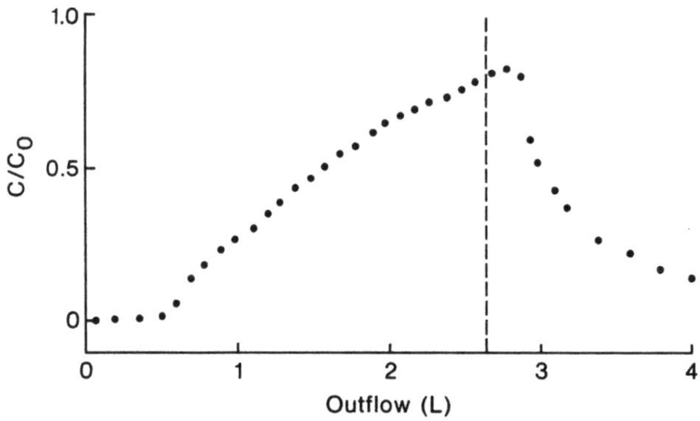

Fig. 5. A breakthrough curve when a phosphate solution entered a Western Australian soil and was subsequently followed by a phosphate-free solution (Aylmore and Murali 1981)

sorption and concentration is made to be curved, the outcome is increased asymmetry (Fig. 4, line 2). If the time component is then added, the asymmetry is further increased because the continuing reaction increases sorption. The effluent concentration therefore increases more slowly but the sorbed material is held more tightly and so desorption is decreased (Fig. 4, line 3). Finally, (line 4) adding the component (k_1) to describe the effects of previous additions on the reaction of the next addition increases the asymmetry still further. This occurs because the ability of the soil to retain each new addition is decreased and so the outflow concentration rises more quickly, yet the retained sorbate is held strongly and so the desorption limb also drops quickly.

The simulated breakthrough curves for which realistic characteristics for sorption were used (line 4, and, to a lesser extent, line 3) seem to reproduce the observed shapes of curves from real experiments, for example those of Aylmore and Murali (1981) (Fig. 5) and those of Mansell et al. (1985). The simulated curves also seem to be closer to the data of Mansell et al. than was the model they used. The simulated curves could have been closely tuned to the observed curves by an appropriate choice of parameters, but this is not a good test of a model. A good test would require a fit both to the components and to the whole. That is, an experiment in which realistic functions are used to describe sorption over an extended period and with repeated additions of reactant. These functions would then be used to predict leaching and these predictions would be tested against observed leaching.

7 Conclusions

Pragmatic sorption and desorption models can be obtained to describe observations over extended periods and a wide range of concentrations. An effective pragmatic model describes sorption as increasing with a small fractional power of time. In this model, desorption is no more than the opposite of sorption of *new* reactant but, of course, it differs from sorption of the original reactant. This model is consistent with a comprehensive, mechanistic model in which the slow process that follows adsorption is assigned to a diffusive penetration of the surface. It is because the rate of this diffusion is modified by electrostatic effects that the rate is approximately proportional to a small fractional power of time. Desorption is slow simply because back diffusion of penetrated material is slow. Neither the pragmatic nor the mechanistic models are compatible with sophisticated descriptions of the movement of water in porous media. The main problem is that the current rate of change of concentration depends on the period that has elapsed since the application was made and on the amounts and history of previous applications. This is an inevitable consequence of the diffusion process involved. However, the models of sorption and desorption are compatable with simple, layer models of the movement of water and these can be used to reproduce characteristic breakthrough curves.

References

Addiscott TM, Wagenet RJ (1985) Concepts of solute leaching in soils: a review of modelling approaches. J Soil Sci 36:411–424

Amacher MC, Kotuby-Amacher J, Selim HM, Iskander IK (1986) Retention and release of metals by soils – evaluation of several models. Geoderma 38:132–154

Aringhieri R, Pardini G (1983) Interaction between OH^- ions and soil surfaces: a kinetic study. Can J Soil Sci 63:741–748

Aringhieri R, Carrai P, Petruzelli G (1985) Kinetics of Cu^{2+} and Cd^{2+} adsorption by an Italian soil. Soil Sci 139:197–204

Atkinson RJ, Posner AM, Quirk JP (1971) Kinetics of heterogenous isotopic exchange reactions: derivation of an Elovich equation. Proc R Soc London Ser A 324:247–255

Aylmore LAG, Murali V (1981) A convective-dispersive-adsorptive flow model for solute transport in soils. II: Evaluation of single and two-component adsorptive models for phosphate movement in soils. Aust J Soil Res 19:287–298

Barrow NJ (1974) Effects of previous additions of phosphate on phosphate adsorption by soils. Soil Sci 118:82–89

Barrow NJ (1978) The description of phosphate adsorption curves. J Soil Sci 29: 447–462

Barrow NJ (1979) The description of desorption of phosphate from soil J Soil Sci 30:259–270

Barrow NJ (1980a) Differences among some North American soils in the rate of reaction with phosphate. J Environ Qual 9:644–648

Barrow NJ (1980b) Differences among a wide-ranging collection of soils in the rate of reaction with phosphate. Aust J Soil Res 18:215–224

Barrow NJ (1983a) A discussion of the methods for measuring the rate of reaction between soil and phosphate. Fertil Res 4:51–61

Barrow NJ (1983b) On the reversibility of phosphate sorption by soils. J Soil Sci 34:751–758

Barrow NJ (1987) Reactions with variable-charge soils. Nijhoff, Dordrecht, Neth

Barrow NJ, Shaw TC (1975a) The slow reactions between soil and anions: 2 Effect of time and temperature on the decrease in phosphate concentration in the soil solution. Soil Sci 119:167–177

Barrow NJ, Shaw TC (1975b) The slow reactions between soil and anions: 4 Effects of time and temperature of contact between soil and molybdate on the uptake of molybdenum by plants and the molybdenum concentration in the soil solution. Soil Sci 119:301–310

Barrow NJ, Shaw TC (1975c) The slow reactions between soil and anions: 5 Effects of period of contact on the desorption of phosphate from soils. Soil Sci 119:311–320

Barrow NJ, Shaw TC (1977a) The slow reactions between soil and anions: 6 Effects of time and temperature of contact on fluoride. Soil Sci 124:265–278

Barrow NJ, Shaw TC (1977b) The slow reactions between soil and anions: 7 Effect of time and temperature of contact between an adsorbing soil and sulfate. Soil Sci 124:347–354

Chien SH, Clayton WR (1980) Applicability of the Elovich equation to the kinetics of phosphate release and sorption in soils. Soil Sci Soc Ame J 40:265–268

Elkhatib EA, Bennet OL, Wright RJ (1984) Kinetics of arsenite sorption by soils. Soil Sci Soc Am J 48:758–762

Elrashidi MA, O'Connor GA (1982) Influence of solution composition on sorption of zinc by soils. Soil Sci Soc Am J 46:1153–1157

Hingston FJ (1981) A review of anion adsorption. In: Anderson MA, Rubin AJ (eds) Adsorption of inorganics at solid-liquid interfaces. Ann Arbor Science Publishers, Ann Arbor, Mich

Ibrahim HS, Pratt PF (1982) Effects of rate of application and time on phosphate sorption by soils. Soil Sci Soc Am J 46:926–928

Kuo S, Lotse EG (1974) Kinetics of phosphate adsorption and desorption by lake sediments. Soil Sci Soc Am Proc 38:50–54

Lin C, Busscher WJ, Douglas LA (1983) Multifactor kinetics of phosphate reactions with minerals in acidic soils: I. Modelling and simulation. Soil Sci Soc Am J 47:1097–1103

Mansell RS, McKenna PJ, Flaig E, Hall M (1985) Phosphate movement in columns of sandy soil from a wastewater-irrigated site. Soil Sci 140:59–68

Novak LT, Petschauer FJ (1979) Kinetics of the reaction between orthophosphate ions and Muskegon dune sand. J Environ Qual 8:312–318

Probert ME, Larsen S (1972) The kinetics of heterogenous isotopic exchange. J Soil Sci 23:76–81

Schwertmann U, Schieck E (1980) Das Verhalten von Phosphat in eisenoxidreichen Kalkgleyen der Münchener Schotterebene. Z Pflanzenernähr Bodenkd 143:391–401

Sibbesen E (1981) Some new equations to describe phosphate sorption by soils. J Soil Sci 32:67–74

Singh BR (1984) Sulfate adsorption by acid forest soils: 4. Kinetics and effects of temperature and moisture. Soil Sci 138:440–447

Tambe KN, Savant NK (1978) Kinetics of sorption of orthophosphate and pyrophosphate by ammoniated tropical soils. Commun Soil Sci Plant Anal 9:745–754

Oxyanion Sorption on Soil Surfaces

U. Kafkafi[1]

1 Introduction

The non-metal elements in the groups IIIA to VIA form oxyanions in water solutions. Some of the inorganic oxyanions commonly found in the aqueous phase in the soil solution are: carbonate, nitrate, silicate, phosphate, selenate and molybdate. The negative charge on the oxyanion is dependent on the pH and the dissociation constant of any specific group. Phosphate, for example, has three OH groups and is present in the solution in all these forms at different proportions as H_3PO_4, $H_2PO_4^-$, HPO_4^{2-} and PO_4^{3-} in the pH range of 0–14. The proportion of each anion form in solution at any specific pH could be calculated from the dissociation constants of each anion and the total amount of P present (Fig. 1). The soil pH can vary within short distances around active living organs like plant roots due to differential absorption of cations (Marschner et al. 1986) or due to respiration that affects the partial CO_2 pressure. Due to the sensitivity of the oxyanion charge to pH and the unpredictable variations in soil pH, abundant literature can be found which deals with the reactions of oxyanions in soils, clays and oxide surfaces.

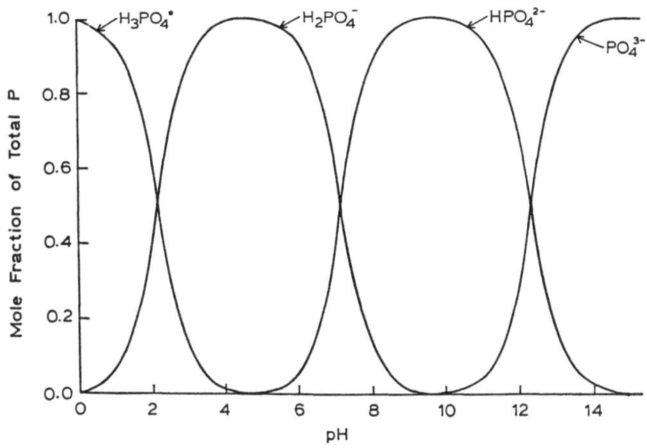

Fig. 1. Distribution diagram for 0.1 M orthophosphate solution $pK_1 = 2.1$; $pK_2 = 7.2$; $pK_3 = 12.3$ (Van Wazer 1958)

[1]Hebrew University of Jerusalem, Dept. of Field Crops, Faculty of Agriculture, Rehovot, Israel

2 Sites of Reactions of Oxyanions

The main reactive sites in soils that are capable of binding oxyanions are the oxides and hydroxides of iron and aluminum. At high pH values, precipitates of Ca^{2+} with oxyanions may create new minerals like apatite. Iron and aluminum oxides exist in the soil in various forms: free ions in solution, metal hydroxy-polymers, with or without oxyanions attached to them in weak or strong bonds. The silicate minerals could be considered as a reaction product between Al-polymers and silicic acid polymers with various degrees of impurities caused by replacing divalent cations in the Al-polymer and trivalent cations in the silicate polymer at various Al/Si ratios. The Fe- and Al-oxides have active surface hydroxyls. These surfaces are usually saturated by the polysilicate layers. In the silicate clay particles the potential reactive sites for the oxyanion are the edge-hydroxylated Al-ions.

The general characteristics of the inorganic surface hydroxyl group can be illustrated (Sposito 1984) by the mineral goethite (α-FeOOH) (Fig. 2). Three types of OH groups are found on the plane perpendicular to the crystallographic *a* axis, denoted as singly (A), triply (B) and doubly (C) coordinated to Fe(III). Type (A) is coordinated to one Fe^{3+} cation in the bulk structure that has become protonated when exposed as a surface cation. Type (C) is the same as (A) but coordinated to two Fe^{3+} cations. Type (B) is an hydroxyl group connected to three Fe^{3+} cations that were exposed to the surface. These hydroxyls could be detected by their specific infrared spectra (Parfitt et al. 1975, 1976). On the plane perpendicular to axis *b* only (C)-type hydroxyls exist, always connected to the Fe^{3+} cation with an accompanying water molecule. The Fe(III)·H_2O arrangement acts as a Lewis acid site where the Fe^{3+} cation is the Lewis acid. The type-(A) hydroxyl group can be protonated and then be exchanged to form a complex with an

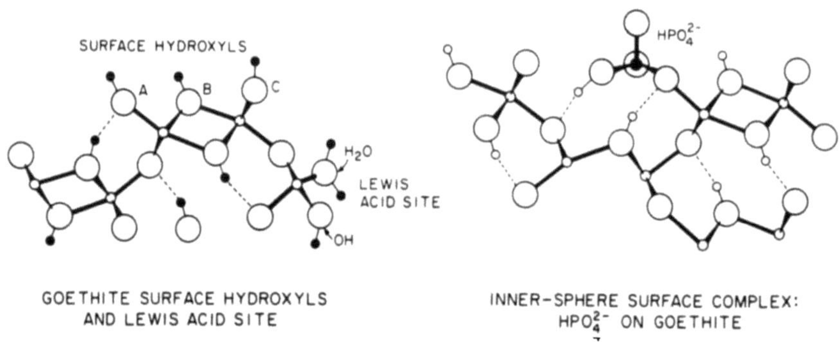

Fig. 2. Surface hydroxyl groups on goethite: singly (A-type), triply (B-type), and doubly (C-type) coordinated to Fe(III), along with Lewis acid site hydroxyls. The drawing on the *right* shows an inner-sphere surface complex with HPO_4^{2-} at the A-type hydroxyl group. The *dashed lines* indicate hydrogen bonds

oxyanion like HPO_4^{2-} (Fig. 2). A similar arrangement exists in gibbsite γ-$Al(OH)_3$ where the Al^{3+} cation, coordinated to a single water molecule, acts as a Lewis acid site. While the basal plane makes up the largest fraction of the surface, with its OH groups connected to pairs of Al^{3+} cations, the Lewis acid sites on the edge are more reactive.

In phyllosilicates, singly coordinated OH groups are created on the edge surface when the crystallites are broken. The gibbsite structure is the backbone of most phyllosilicates, which are essentially polysilicate layers adsorbed on the planar surface of the gibbsite, leaving the edge OH groups as the reactive sites for other oxyanions present in the solution.

3 Inorganic Oxyanion Adsorption

It is accepted that an oxyanion adsorption by soil mineral involves a two-step ligand exchange reaction (Sposito 1984).

$$SOH_{(s)} + H^+ (aq) = SOH_2^+ (s) \quad (1)$$
$$SOH_2{}^+{}_{(s)} + L^{z-}{}_{(aq)} = SL^{1-z}{}_{(s)} + H_2O \quad (2)$$

where S is the metal cation, $SOH_{(s)}$ is 1 mol of the surface hydroxyl group and L^{z-} is an oxyanion of valence z.

The protonation step (1) is suggested to render the SOH more exchangeable at low oxyanion concentrations; this step could be avoided at a high concentration of the oxyanion (White 1981). One of the earlier works that demonstrated the formation of a free water molecule during adsorption of phosphate on kaolinite was published in 1939 (Stout).

The schematic four layer adsorption model (Fig. 3) offers a molecular description of surface complexation reactions. As can be seen in Fig. 3 the crystal-surface hydroxyl plane, marked “s”, represents the sphere that can be penetrated by oxyanions which can contribute at least one OH group to the bond. Due to the intimate and close association with the surface metal cation, the adsorbed species, the center of which runs through plane “a”, is referred to as the “inner-sphere” (Sposito 1984) or “inner Helmholtz plane” (Hingston et al. 1968) complexes.

The example of the phosphate ion in Fig. 3 represents a possibility that the oxyanion ligand is complexed with two Al^{3+} at the surface through an oxygen atom. This formation is possible since the crystal distance between these two oxygen atoms is very close to the distance between the oxygen atoms at the base of the phosphate tetrahedron (Kafkafi et al. 1967). When a high concentration of the P ligand is present in solution, a single Al-O-P bond is established. The H on the OH group of the adsorbed phosphate tetrahedron is easily replaced by other monovalent cations. This behaviour partly explaining the well-known fact that phosphate adsorption increases the cation exchange capacity of oxisols.

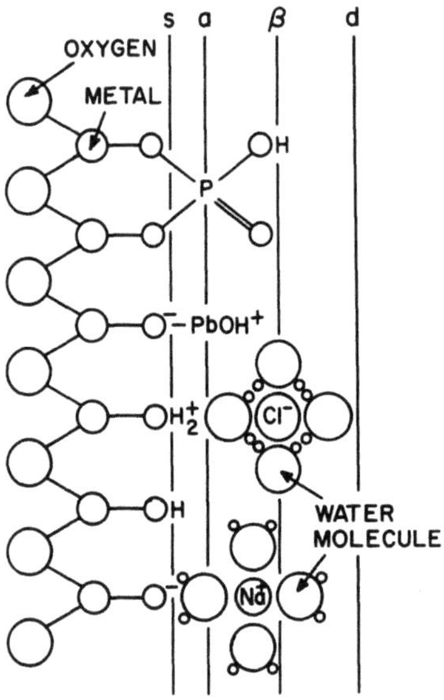

Fig. 3. A schematic portrayal of an inorganic hydroxyl surface, showing planes associated with surface hydroxyl groups ("s"), inner-sphere complexes ("a"), outer-sphere complexes ("β"), and the diffuse ion swarm ("d") (Sposito 1984)

Ions like Cl^- or NO_3^- surrounded with water molecules can approach a protonated surface at a distance of at least one water molecule. They are referred to as "outer-sphere" complexes "β". The plane "d" is referred to as the diffuse ion swarm.

The assumptions of the multi layer model, according to Sposito (1984), are:

1. Only H^+ and OH^- can form "inner-surface" complexes; all other adsorbed metal cations and inorganic and organic anions remain in the "outer sphere".
2. The reacting surface functional group is OH. When anionic complexes are adsorbed, the equation obeyed is:

$$bSOH_{(s)} + qL^{l-}_{(aq)} + xH^+_{(aq)} = S_bH_xL^{\gamma}_{q(s)} + bOH^-_{(aq)}, \quad (3)$$

when $\gamma = x + b - ql$.

3. The relationship between surface charge δ and the inner potential is specified through the capacitance equations:

$$\sigma_H = C_1(\Psi_s - \Psi_\beta); \quad (4)$$
$$\sigma_d = C_2(\Psi_d - \Psi_\beta); \quad (5)$$

where C_1 and C_2 are integral capacitance densities and the inner potentials Ψ_s, Ψ_β and Ψ_d are identified with the three planes "s", "β" and "d". The ligand exchange approach of oxyanion adsorption is supported by many experimental observations.

1. The dependence on pH. Under a set of conditions the oxyanion adsorption is controlled by the pH of the experimental slurry (Fig. 4). When one protonated

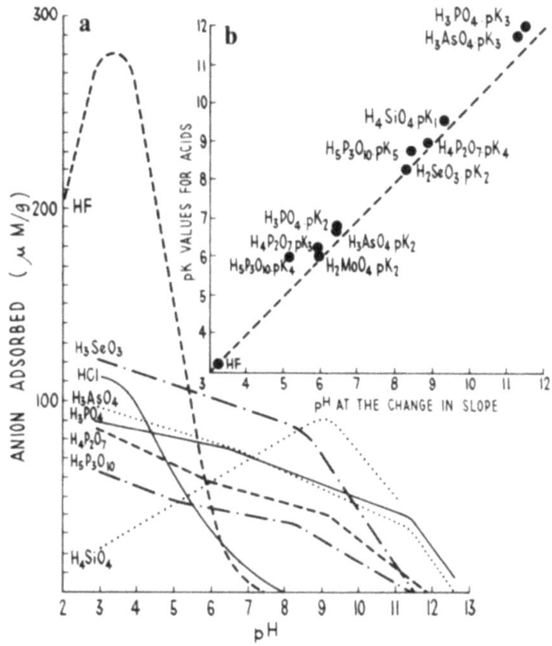

Fig. 4a. Adsorption "envelopes" for anions with 0.1 M NaCl as supporting electrolyte. The curve for Cl- illustrates non-specific adsorption. **b** The plot of pKs for weak acids against pH for breaks in slope of adsorption "envelopes." The *broken line* indicates the ideal relationship. Except pK_3 for H_3PO_4 and H_3AsO_4 dissociation constants were determined in 0.1 M NaCl. Approximate values of the exceptions were obtained from values given by Sillen (1964). The values for molybdate were taken from adsorption data given by Reisenauer, Tabikh and Stout (1962) (Hingston et al. 1968)

species is available, as in the case for silica in the pH range of 5–10, a peak in the adsorption is obtained near the pK_a of the ligand acid. When more than one species exists within the experimental pH range, the adsorption envelope (maximal adsorption versus pH value) usually shows a decline with an increase in pH but with changes in slope at the corresponding pK_a of the ligand species. Correlation of adsorption maxima and pK values of the ligand give a linear relationship (Fig. 4; Hingston et al. 1968, 1972).

2. The kinetics of P^{32} exchange. At a relatively high P concentration the rate of isotopic exchange follows a first-order dependence on the adsorbed phosphate concentration. This rate-limiting step was interpreted as the need to break either the metal-oxygen bond in the adsorbent or the metal phosphate bond in the inner-sphere surface complex (Atkinson et al. 1972).

3. Reversibility of adsorption in the presence of indifferent electrolytes: Oxyanions on clay surfaces show hysteresis with respect to reduction in ligand concentration. Kafkafi et al. (1967) has shown on kaolinite surfaces that the "equilibrium" concentration of phosphate in the solution is not a unique value of surface-adsorbed P. For the same P concentration in the solution many values of adsorbed P can be obtained, depending on the way of approaching this point (Kafkafi et al. 1967). Based on isotopic exchange they have postulated a double bond of a P tetrahedron to two-edge Al through an oxygen to create a six-membered ring with a stable chair formation. Later, Parfitt et al. (1975) observed

the same phenomena on iron oxides and were able to support that type of bond by IR spectroscopy. The reversibility of P adsorption with respect to its concentration only could not be obtained when the fraction of P on the surface is low. At very high concentrations (Kafkafi unpublished results) of P all the adsorbed P becomes exchangeable, indicating that the double bond to the surface was broken by competing P in the solution.

The ability of other oxyanions to compete with the adsorbed P *is well known*. It has been long suggested to use citric acid (Dyer 1894) or bicarbonate ion (Olsen et al. 1954) to replace adsorbed P and to estimate its availability to plants.

Studying the effects of ionic strength, pH, slurry density and competition of organic oxyanions with P adsorption on kaolinite and montmorillonite monoionic clays (Kafkafi et al. 1988), it was observed that increasing ionic strength of the indifferent Cl^- ion in the solution increased the amount of adsorbed P. The maximum adsorption was obtained at pH 5.1 for kaolinite and at ~ 6 for montmorillonite at the PZC of the clays. Using the competitive adsorption model (Bar-Yosef 1979) the experimental data in Fig. 5 cannot be described by the model that considers only an affinity-constant, ionic activity of all the ions including the OH and other competing ions. This model assumes that there is a constant maximum adsorption site T for the adsorption. This assumption may hold for analytically prepared oxide surfaces, it has been shown for silicate minerals and soils (Kafkafi and Bar-Yosef 1969) that silicate ions leave the clays when long equilibration in solution is allowed and that this process is enhanced by phosphate even at very low concentrations. It is therefore possible that the number of maximum adsorption sites is not constant on silicate minerals and that it changes with the experimental conditions.

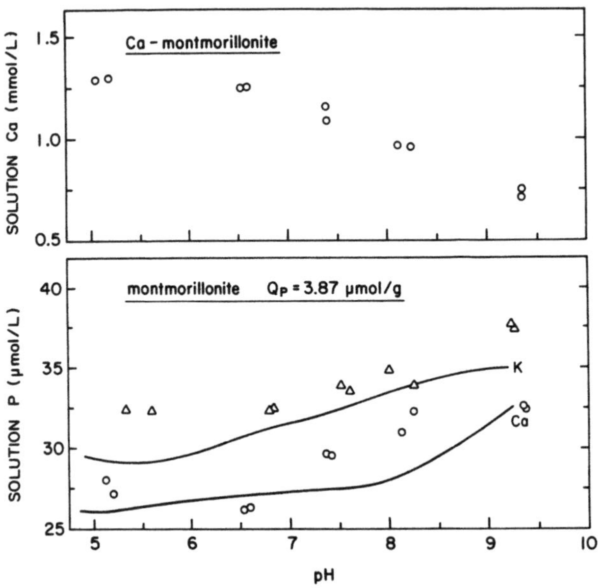

Fig. 5. P concentration in the solution of K and Ca-montmorillonite suspensions as a function of suspension pH. Electrolyte concentration 0.03 M KCl; constant amount of P (Qp). Slurry density – 1 g/100 g. The *solid line* represents a computed competitive model (Bar-Yosef et al. 1988). The Ca concentration in each pH is given at the top (Bar-Yosef et al. 1988)

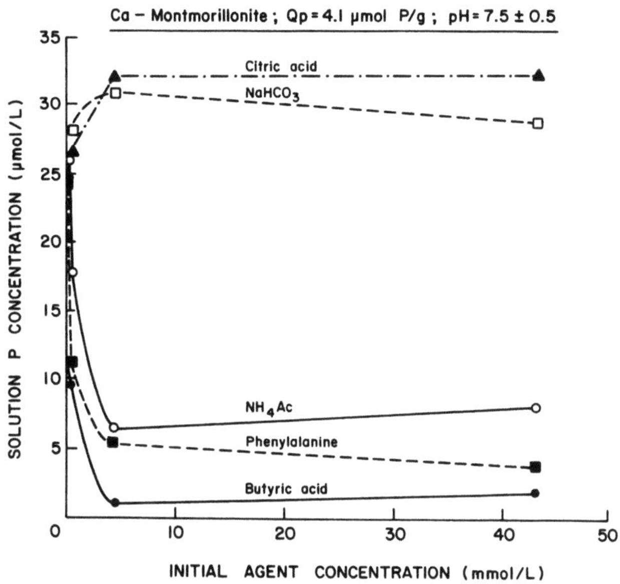

Fig. 6. Effect of organic agent concentration on desorption of previously adsorbed phosphate (Qp) on Ca-montmorillonite (Kafkafi et al. 1988)

The effect of the organic oxyanions was different (Fig. 6). The carboxyl group in bicarbonate and the citrate ions competed and removed the P from the clay, while the carboxylic groups in α-amino isobutyric acid and L-phenylalanine as well as in the ammonium acetate increased the amount of P adsorbed to the surface. The unifying principle that explains these findings (Kafkafi et al. 1988) is that P adsorption on a clay depends on the extent of the depression of the negative potential of the clay platelets. Bicarbonate and citrate anions at $pH > 8$ compete with P for active Al-edge sites on the clay. The increased P adsorption in the presence of high ionic strength and amino acids with the carboxylic group is suggested to stem from the fact that the amino group is very effective in producing H-bonding causing the collapse of the interlamellar spacing.

When the single platelet of montmorillonite is present in dilute salt solutions, which represents the general working conditions in adsorption studies, the negative force lines emanating from the negatively charged surface spill over the edge (Secor and Radke 1985) and prevent the approach of the negatively charged oxyanion to the Al site of the edge of the crystal lattice. When several platelets collapse and create a quasi-crystal, the negatively electrostatic field force lines spilling over the platelet edge do not cover the whole stacked edge, leaving free access to active Al sites on the clay (Fig. 7).

The conclusion that Al active sites at the broken edges of silicate minerals are the adsorption sites for phosphate and probably other oxyanions was reached earlier by Goldsztaub et al. (1954) and Pissarides et al. (1968). Keren and O'Connor (1982) have postulated such a bond for boron on illite and montmorillonite.

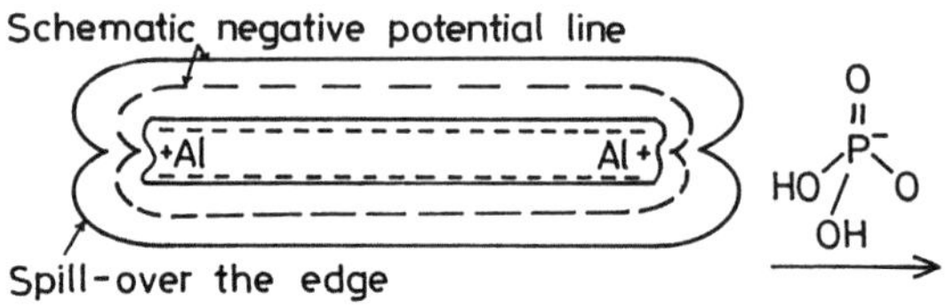

SINGLE DISPERSSED MONTMORILLONITE PLATELET: Negative P adsorption. Phosphate ion approach to Al (+) site is blocked by spill-over of negative potential.

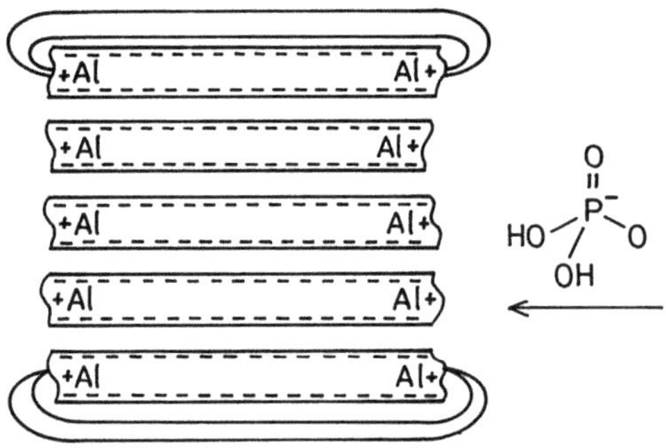

QUASICRYSTAL OF MONTMORILLONITE: Positive P adsorption. Phosphate ion is free to approach active Al (+) site on edges of clay. Spill-over of negative potential affects only small portion of the edge surface.

Fig. 7. Descriptive model of adsorbed P on a quasi-crystal of montmorillonite

Decreasing the water content of the slurry below 10 l kg^{-1} (Fig. 8) also increased the amount of adsorbed P on the surface of Ca-montmorillonite (Bar-Yosef et al. 1988). Since under actual soil conditions the water/clay ratio is even smaller, serious doubt exists about the validity of extrapolating conclusions on the oxyanion behaviour in the actual field from dilute solution studies. The early method of water displacement from soil columns to obtain a relevant soil solution composition (Burd and Martin 1923) should be recalled especially with regards to the studies of the surface-soil solution distribution of oxyanions.

The dependence of the amount of oxyanion adsorption on the pH, ionic strength, amount of exposed active Al-sites and the presence of other competing oxyanions make it very difficult to predict the relationship between the adsorbed species and its concentration in the soil solution. Remembering the fact that under natural conditions biological activity of plant roots and rhizophere may create a pH change of more than ± 2 units in a very short time, it is no wonder that empirical studies are still needed to answer practical agricultural and environmental problems.

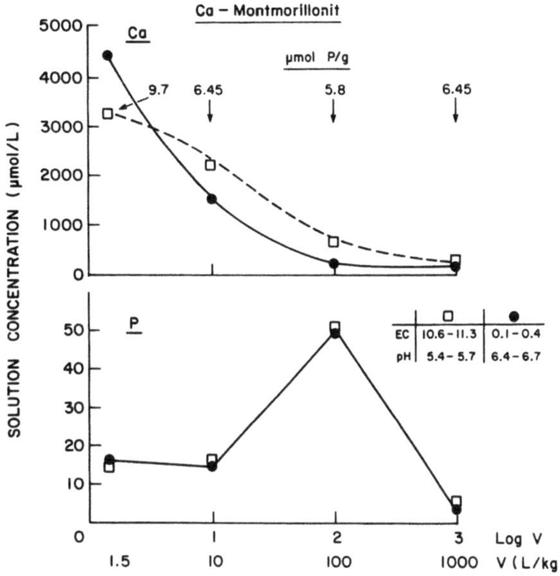

Fig. 8. Effect of clay slurry density of two pH and EC values on P and Ca concentrations in the equilibrium solution of Ca-montmorillonite. *Numbers* above *arrows* indicate the total quantity of P, μmol P g^{-1} at each (Kafkafi et al. 1988)

Our understanding of the reaction mechanisms and bonding model help us to anticipate the general directions of oxyanion reaction.

This knowledge is of great help in devising practical field activities even if the exact quantitative relationships between the adsorbed and solution ligand could not be achieved at the present stage of knowledge.

References

Atkinson RJ, Posner AM, JP Quirk (1972) Kinetics of isotopic exchange of phosphate at the α-FeOOH-aqueous solution interface. J Inorg Nucl Chem 34:2201

Bar-Yosef B (1979) pH Dependent Zn adsorption by soils. Soil Sci Soc Am J 43:1095–1099

Bar-Yosef B, Kafkafi U, Rosenberg R, Sposito G (1989) Competitive adsorption of P by Ca- and K-saturated kaolinite and montmorillonite: 1. Effect of equilibration time, ionic strength, P concentration and pH. Soil Sci Soc Am J 52:1580–1585

Burd JS, Martin JC (1923) Water displacement of soils and soil solution. J Agric Sci 13:265–295

Dyer B (1894) On the analytical determination of probable available mineral plant food in soils. Trans Chem Soc 64:115–167

Goldsztaub S, Henin S, Wey R (1954) Sur l'adsorption d'ions phosphoriques par les argiles. Clay Mineral Bull 2:162–166

Hingston FJ, Atkinson RJ, Posner AM, Quirk JP (1968) Specific adsorption of anions on goethite. 9th Int Congr Soil Sci Adelaide, Aust, 1968, pp 669–678

Hingston FJ, Posner AM, Quirk JP (1972) Anion adsorption by goethite and gibbsite: I. The role of the proton in determining adsorption envelopes. J Soil Sci 23:177–192

Kafkafi U, Bar-Yosef B (1969) The effect of pH on the adsorption and desorption of silica and phosphate on and from kaolinite. Proc Int Clay Conf, Tokyo. Israel Univ Press, Jerusalem,pp 691–696

Kafkafi U, Posner AM, Quirk JP (1967) Desorption of P from kaolinite. Soil Sci Soc Am Proc 31:348–353

Kafkafi U, Bar-Yosef B, Rosenberg R, Sposito G (1988) Competitive adsorption of P by Ca- and K-saturated kaolinite and montmorillonite: II. Organic anion competition. Soil Sci Soc Am J 52:1585–1589

Keren R, O'Conner GA (1982) Effect of exchangeable ions and ionic strength on boron adsorption by montmorillonite and illite. Clays Clay Minerals 30:341–346

Marschner H, Romheld V, Horst WJ, Martin P (1986) Root induced changes in the rhizosphere: importance for the mineral nutrition of plants. Z Pflanzenernähr Bodenkd 149:441–456

Olsen SR, Cole CV, Watanabe FS, Dean LA (1954) Estimation of available phosphorus in soil by extraction with sodium bicarbonate. US Dep Agric Circ 939

Parfitt RL, Atkinson RJ, St C Smart R (1975) The mechanism of phosphate fixation by iron oxides. Soil Sci Soc Am Proc 39:837–841

Parfitt RL, Russell JD, Farmer VC (1976) Confirmation of the surface structure of goethite(α-FeOOH) and phosphated goethite by infrared spectroscopy. JCS Farady I 72:1082

Pissarides A, Stewart JWB, Rennie DA (1968) Influence of cation saturation on phosphorus adsorption by selected clay minerals. Can J Soil Sci 48:151–157

Secor RB, Radke CJ (1985) Spillover of the diffuse double layer on montmorillonite particles. J Coll Interface Sci 103:237–244

Sposito G (1984) The surface chemistry of soils. Oxford Univ Press, New York

Stout PR (1939) Alteration in the crystal structure of clay minerals as a result of phosphate fixation. Soil Sci Soc Am Proc 4:177–182

Van Wazer Jr (1958) Phosphorus and its compounds, vol 1. Chemistry. Interscience, New York

White RE (1981) Retention and release of phosphate by soil and soil constituents. In: Tinker PB (ed) Soils and agriculture. John Wiley & Sons, New York

The Effects of Micro-Scale Solute Movement on Biotic Transformations in Porous Media

P.R. Darrah[1]

1 Introduction

The variety of agrochemicals, fertilizers, sewage sludge and other potentially toxic products currently being applied to soil in ever increasing quantities is of great concern to scientists, environmentalists, public health and water authorities and the public at large. Arising from this concern, there is a need for reliable quantitative methods for predicting the behaviour of these potential pollutants in the soil in terms of their environmental impact on soil ecosystems, their persistence in the soil environment and their rate of leaching to groundwater. This requirement for predictive capacity has led to an inexorable rise in the number of mathematical simulation models which, to differing degrees, attempt to produce some quantitative prediction of the likely behaviour of a pollutant in a particular soil environment.

At the same time, there has been an increasing recognition of the role of the soil biota, especially the micro-biota, in determining or modifying the fate of soil pollutants and many simulation models must now include some biological components in their formulation if realistic predictions are to be made.

If a mathematical model is to have any value as a management tool, then it must be capable of predicting the behaviour of pollutants over a wide range of soil and environmental conditions; however, for most models of soil biological processes this does not seem to be the case. The purpose of this chapter is to identify those processes which seem to be important in determining the biological activity of soils, but which have been neglected in the formulation of most published simulation models.

2 Types of Model

Models vary widely in their aims, scope and degree of complexity, often reflecting the expertise and particular interests of the modeler. It is useful to distinguish between the various types of model and two classes (or more correctly continua) are commonly used.

[1]Department of Plant Sciences, University of Oxford, Agricultural Science Building, Parks Road, Oxford OX1 3PF, England

1. Models may be classed as either deterministic or stochastic. A deterministic model is one in which a unique outcome is predicted for a specified system and for a given set of initial conditions. A stochastic model recognizes the inherent variability of natural systems and the consequent difficulty of defining the system adequately or of specifying the initial conditions. A stochastic model therefore presupposes the outcome of any modelling exercise to be uncertain and predicts the probabilistic outcome. The stochastic approach is of fairly recent origin in the modelling of soil processes and has been largely confined to the prediction of water and solute movement in soil (Addiscott and Wagenet 1985; Jury et al. 1986).

2. The second classification continua are based on the extent to which fundamental mechanisms are used in the formulation of the model. At one extreme lies the functional or empirical model which is essentially a method of describing experimental data by a mathematical expression. Empirical models represent a powerful tool for reducing large and variable sets of data to more manageable forms and, by their very nature, have a very high predictive capacity over the limited range of the experimental data on which they were based. At the other extreme of this continuum lies the fully mechanistic model, in which each process considered by the model is treated with as much insight into the fundamental mechanisms as is available or as the modeler considers is warranted. By definition, therefore, a purely mechanistic model is unattainable because the resolution of the mechanistic treatment can always be improved; for instance, most 'mechanistic' models of microbial growth in soil use the Monod equation to describe growth, however, the Monod formalism is itself empirical in that it does not consider the individual enzyme reactions which contribute to the growth of the organism. Perhaps a more useful definition of a mechanistic model is that it is a series of assumptions or hypotheses about how a particular process or system operates, cast in mathematical form. This definition also illustrates the essential feature of any mechanistic model; that it is hypothesis-based and therefore in common with all scientific hypotheses requires experimental testing before it can be considered to have any validity or predictive power. Regrettably, the literature abounds with models which have undergone no or a very limited form of testing and whose predictions therefore must be regarded as purely speculative. The power of a properly validated mechanistic model is its capacity to predict the outcome of experiments a priori provided that the various input parameters are known. Deterministic, mechanistic models are the most common form for research and increasingly for management purposes and this review concentrates on this class of model.

3 Mechanistic Models of Soil Processes

Soil is an extremely heterogeneous medium at all levels of observation. At the microscopic level, heterogeneity is observed in the type and arrangement of primary minerals with different physical, chemical and biological properties.

Expanding the scale of observation reveals the organization of the soil minerals into micro-aggregates, aggregates, etc. and associated with this level of observation is the property of porosity between and within aggregates. Further scale expansions reveal a vertical, spatial heterogeneity due to profile development.

At every level of observation, two components of this heterogeneity can be recognized, systematic (or deterministic) variation and random (or stochastic) variation (Wilding and Drees 1978; Philip 1980). Deterministic variation is characterized by a structure to the variability in a particular soil property, e.g. the variation displays a spatial or temporal correlation. Deterministic variation is also a feature which is explicable mechanistically in terms of the action of soil processes (Wilding and Drees 1978). Random variation or 'noise', on the other hand, displays no structure in the pattern of its variation. Inherent in the distinction between systematic and random variation therefore is that the former may be predicted from a knowledge of the various soil processes involved, while the latter is essentially unpredictable. Mechanistic, deterministic models can only predict *changes* in the systematic variation in the soil since, by definition, they cannot predict *changes* in the random component of variation. However, the distinction between systematic and random variation is not absolute, but rather depends on the scale on which the observations are made: thus, if the scale is enlarged then variation previously classed as random will be seen to possess structure and therefore to be deterministic (Burrough 1983). This reclassification of random to systematic variation with enlargement of the observational scale is associated with the idea that different soil processes contribute to systematic spatial variation over different spatial scales (Fridland 1976).

It would be reasonable to expect that mechanistic models of soil processes would reflect in their formulation the various spatial scales over which soil processes operate. However, in the main, this is not the case: most published mechanistic models consider only one spatial scale and they class the systematic variation apparent at larger scales as totally random variation which must be accounted for by using a single value for the particular soil property which is averaged over a finite volume of soil.

As a typical example, consider the model of urea diffusion and degradation developed by Rachhpal-Singh and Nye (1984). In this experiment, urea was applied uniformly to the surface of a 4-cm diameter column of soil and, after allowing time for diffusion and microbial degradation, the column was sectioned and the urea content of each section determined. The simulation model they developed was designed to predict the distribution of urea along the vertical axis of diffusion. The different spatial scales implicit in the experimental system should be noted: the diffusive transport of urea with soil depth can be considered as a macro-scale process while the diffusion and degradation of urea within individual soil aggregates is a micro-scale process. In the model, any systematic variation which may have existed in the horizontal dimension at the scale of individual soil aggregates was ignored and the assumption was made that the soil could be regarded as a homogeneous medium along the horizontal axis. The kinetics of urea hydrolysis were described by reference to macro-scale parameters. Thus, the dependence of enzyme activity on urea concentration was described by reference to the macro-scale or bulk concentration and not by reference to the micro-scale concentration in the environment of the enzyme. The simulation model was therefore a one-scale macroscopic model. The authors found excellent fits between their experimental results and model simulations.

It is tempting to conclude that the success of the above model and the many similar models developed for other purposes validates the one-scale macroscopic approach. This review attempts to demonstrate that such a conclusion may not always be prudent and that the effect of ignoring micro-scale processes in the formulation of macro-scale biological models may lead to the development of models of very limited scope, which are closer to empirical models than their mechanistic formulation would suggest.

4 Micro-Scale Effects on Biotic Transformations

The interchange of materials between an organism and its environment only occurs at the interface between them, i.e. organisms only have access to materials from their immediate micro-environment. Therefore, the uptake of material by an organism must deplete the local environment of the material. If the uptake process is to continue, this depleted material must be replaced by transport of the material from elsewhere. Because all organisms must continually exchange materials with their environment if they are to survive, it follows that all biological processes must depend on transport processes occurring externally to the organism as well as on intracellular metabolic processes. It also follows that if the inherent capacity of an organism to remove material from its immediate surroundings exceeds the capacity of the larger environment to replace it, then these external transport processes will limit cellular functioning.

Even in apparently simple situations the role of external transport processes cannot always be ignored. For instance, consider the case of the uptake of solutes by a cell suspension from a well-stirred solution. In this situation, each cell would be surrounded by an unstirred layer of solution across which substrates have to diffuse to reach the cell surface. This case was discussed by Dietschy (1978) and is shown diagrammatically for part of the cell membrane in Fig. 1. It is assumed that the uptake of substrate from the substrate pool immediately surrounding the cell can be described by an equation of the form:

$$F_i = kC_2, \qquad (1)$$

where F_i is the rate of uptake across the unit surface area of the membrane, k is the rate constant and C_2 is the concentration of substrate immediately surrounding the cell.

The rate of diffusion across the unstirred layer surrounding the cell is given by Fick's first law:

$$F_e = -D\,\frac{dc}{dx}, \qquad (2)$$

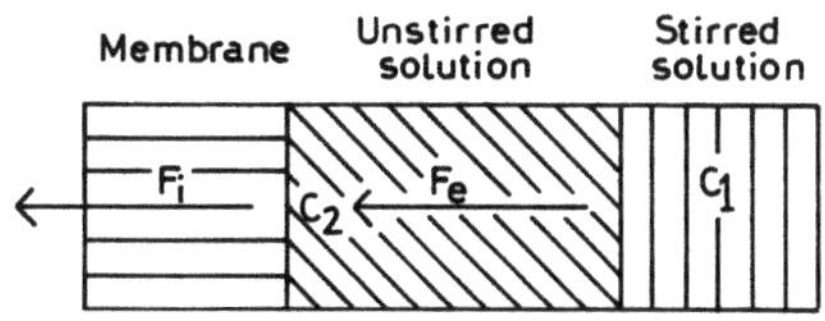

Fig. 1. A diagrammatical representation of the diffusive flux of solute from the bulk stirred solution through the unstirred layer to the surface of a biological membrane, *Fe* and the rate of uptake by the membrane, F_1

where F_e is the flux of substrate across the unstirred layer and D is the diffusion coefficient. This equation may be replaced by a finite difference approximation (Smith 1978):

$$F_e = D\,\frac{(C_1 - C_2)}{\Delta x}\,. \tag{3}$$

The concentration of substrate at the cell surface is not known but Eq. (3) can be used to eliminate C_2 from Eq. (1) and to cast the equation in terms of the known concentration of substrate in the well-stirred bulk solution, C_1:

$$F_i = \frac{k(C_2 - F_e D)}{\Delta x}\,. \tag{4}$$

At steady state, $F_1 = F_e$, and therefore

$$F_i = \frac{kC_1}{\left(1 + \dfrac{kD}{x}\right)}\,. \tag{5}$$

It is apparent from Eq. (4) that even in well-stirred, simple biological systems, the rate of substrate uptake may be limited by the diffusion of the substrates through the external environment. If such a system was being modelled using the macro-scale assumption outlined above, these small-scale transport processes would be ignored. In this case, changes in the experimental conditions which only altered the external diffusion characteristics of the substrate, would change the rate of uptake observed experimentally, while no such changes would be predicted by the model. Hence, the scope of the macroscopic model would be limited to a narrow range of experimental conditions, a feature which is typical of an empirical rather than a mechanistic model.

5 Micro-Scale Transport Processes in Artificial Porous Media

Techniques for immobilizing enzymes and, more recently, whole cells on porous supports are widely used in the synthesis of compounds with industrial and medical applications. Many types of support are used including agarose, polyacrylamides and other organic polymers, inorganic oxides and porous glass (Zaborsky 1973). Such commercial applications have stimulated extensive research into the physical, chemical and biological properties of such systems. Much of this research is directly relevant to the study of biological processes in soil.

As an example, consider the simple case of an isolated porous spherical particle, throughout which an immobilised enzyme is distributed uniformly and which is immersed in a well-stirred solution of the enzyme substrate. This case was analyzed by Engasser and Horvath (1973) and the microscopic mechanisms which they considered were the diffusion of substrate into the porous support from the bulk external substrate pool and the simultaneous catalysis of the substrate within the support. In this system, it is apparent that not all the enzyme within the support will experience the same substrate concentration: enzyme close to the centre of the sphere will experience a lower concentration because some substrate will have been transformed by enzyme in outer shells of the sphere as it diffused through the support.

For steady state conditions, the mathematical representation of the system can be written as:

$$D_p \left(\frac{\partial^2 S}{\partial r^2} + \frac{2}{r} \frac{\partial s}{\partial r} \right) = V_{max} \frac{S}{(S + Km)}, \tag{6}$$

where D_p is the effective diffusion coefficient in the porous support, S is the substrate concentration and r the radial position in the sphere. The right-hand side of Eq. (6) corresponds to the Michaelis-Menten expression for enzyme activity in free solution. By defining the dimensionless quantities $\alpha = r/R$ (where R is the radius of the sphere) and $\beta = S/Km$, Eq. (6) can be written in the form:

$$\frac{\partial^2 \beta}{\partial \alpha^2} + \frac{2}{\alpha} \frac{\partial \beta}{\partial \alpha} = R^2 [V_{max}/(Km\, D_p)] \frac{\beta}{(1 + \beta)}. \tag{7}$$

From this equation, a single parameter termed the modulus, ϕ, can be defined as:

$$\phi = \frac{R}{3} \left[\frac{V_{max}}{Km\, D_p} \right]^{\frac{1}{2}}. \tag{8}$$

From inspection of this modulus, it is apparent that the steady state concentration gradient of substrate within the porous support will depend on three factors: the radius of the porous sphere, the rate of substrate diffusion within the sphere and the intrinsic activity of the enzyme within the sphere. The effect of changes in the value of ϕ on the predicted intra-sphere concentration profile of substrate is shown in Fig. 2. Figure 2 shows that the effect of increasing the modulus is to lower the substrate concentration experienced by the enzyme molecules at all points in the sphere and for $\phi > 2$, it is only enzyme located in the outer shell of the support that is actually exposed to substrate and participates in the reaction. Because of this gradient of substrate concentration within the support, the overall rate of reaction is lower than it would be in the absence of diffusional limitation. This is illustrated in Fig. 3 which shows the overall reaction rate as a function of the dimensionless substrate concentration in the external solution for different values of the modulus ϕ.

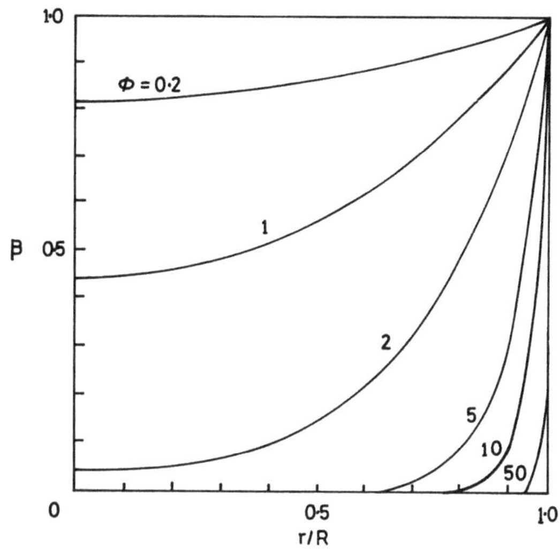

Fig. 2. The concentration profiles of a substrate in spherical immobilized enzyme particles. The dimensionless substrate concentration, β, is plotted as a function of the radial position in the sphere for different values of the modulus, ϕ

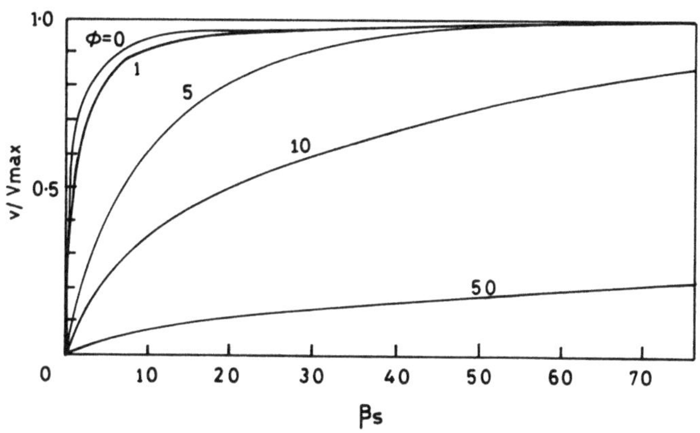

Fig. 3. The dimensionless overall rate of reaction of an immobilized enzyme within a spherical particle as a function of the dimensionless substrate concentration at the surface of the sphere, β_s, with the modulus, ϕ as the parameter

It is apparent from this simple treatment that micro-scale transport limitations in the supply of substrate can have profound effects on the rate of biological processes in porous media. A similar argument can be used to model the situation for immobilized cells, although this treatment is complicated by growth of the cells in the support (Mattiasson 1983). The argument can also be extended to cases where electrostatic interactions occur between the substrate (or the enzyme products) and the support (Engasser and Horvath 1975) or where the enzyme is not uniformly distributed throughout the support (Pedersen et al. 1985), as would be the case in many soil systems.

Although these immobilized enzyme systems are much simpler than any soil system, consisting as they do of regular-sized particles of uniform porosity and with well-defined enzyme loadings, they are nevertheless relevant to the situation in soil and are useful for identifying situations in the soil where similar diffusional limitations may be important.

6 Micro-Scale Diffusion of Oxygen in Soil

Anaerobiosis will occur at any point in the soil where the rate of oxygen consumption by biotic aerobic respiration exceeds the rate of supply of oxygen from the atmosphere. Conditions which favour anaerobiosis are therefore an abundant supply of readily available C to support respiration and high moisture contents, which both favour microbial activity and reduce the effective diffusion coefficient of oxygen through the soil. (The diffusion coefficient of O_2 through a dry soil is four orders of magnitude higher than that in wet soil.) Greenwood and Berry (1962) were among the first workers to recognize that such conditions could occur on the micro-scale, where the rate of supply of O_2 into water-saturated respiring soil aggregates could be sufficiently limiting to allow anoxic conditions to develop in the centre of aggregates of only a few mm radius. However, both this model and a subsequent model of the distribution of oxygen with depth in a water-saturated profile (Radford and Greenwood, 1970) were single-scale models which differed only in their geometrical representation of the soil system and which are directly comparable through the use of scale factors (Tillotson and Nielsen 1984).

A two-scale model was developed by Smith (1980), which considered the diffusion of O_2 into an unsaturated soil profile. The two-scale components were (1) the diffusion of the gas down the soil profile, primarily through the air-filled inter-aggregate pores which was treated as a macroscopic phenomenon and (2) the diffusion of O_2 from the inter-aggregate pore space into respiring soil aggregates which was treated as a micro-scale process. Smith showed that the earlier one-scale model and his two-scale model differed in their predictions of the distribution of anoxic regions in the soil. In the one-scale macroscopic model, oxygen penetration into the profile was fairly limited, but it was also predicted that the entire soil profile above the depth at which O_2 disappeared would be aerobic. The two-stage model, however, predicted a much greater penetration of the gas into the profile, but showed that even at the soil surface approximately 65% of the total aggregate volume would be anaerobic. Such differences would have profound effects on the prediction of an associated process such as denitrification and pesticide fermentation (Tiedje et al. 1984) as the occurrence of anoxic micro-sites close to the soil surface would favour these processes.

This model also illustrates the effect of scale on spatial variability in that the occurrence of anaerobic micro-sites at the centre of soil aggregates would be classed as random variation when viewed on a macroscopic scale, but would appear as structured gradients of O_2 concentration within individual aggregates when viewed on a micro-scale.

7 Micro-Scale Diffusion of Enzyme Substrates in Soil

Most experiments concerned with the kinetics of enzymes in soil have been interpreted using the Michaelis- Menten equation which relates the initial rate of substrate hydrolysis (v) to the substrate concentration (S) i.e.

$$v = V_{max} \frac{S}{S + Km.}, \tag{9}$$

where V_{max} is the maximum rate of substrate hydrolysis and Km is the saturation constant. The substrate concentration has been taken to be the concentration in the bulk soil solution (for a substrate with no electrostatic interactions with the soil solid phase). This equation has been shown to be widely applicable to the initial rate-substrate relations of a wide range of soil enzymes, although attention has been drawn to some cases of poor fits (Irving and Cosgrove 1976). Perhaps because of the almost universal applicability of the Michaelis-Menten equation to describe the kinetics of soil enzymes, the possibility of micro-scale diffusion limitations of the type found in artificial, porous media has not generally been considered.

More recently, it has been shown (Darrah 1982) that, while the Michaelis-Menten equation can be used to fit initial rate data, the kinetic constants obtained may not be constant but may vary with the experimental conditions used. In an assay of a soil β-glucosidase (Darrah and Harris 1986), it was found that the initial rate of hydrolysis by the enzyme at a particular substrate concentration depended not only on the concentration of substrate in the bulk solution but also on the size of the soil aggregates used in the assay. Thus, in a series of assays covering aggregate fractions with mean equivalent sphere radii of 9.3–500 μm, it was found that while V_{max} remained approximately constant, the affinity constant, Km, depended on the mean aggregate size used in the assay for aggregates > 64 μm (Fig. 4). In a further experiment, they found that when different amounts of

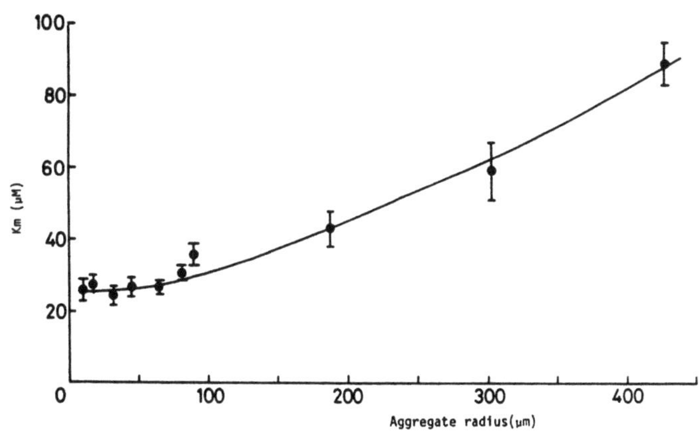

Fig. 4. The variation in Km with soil aggregate radius for soil aggregates which were maintained in suspension

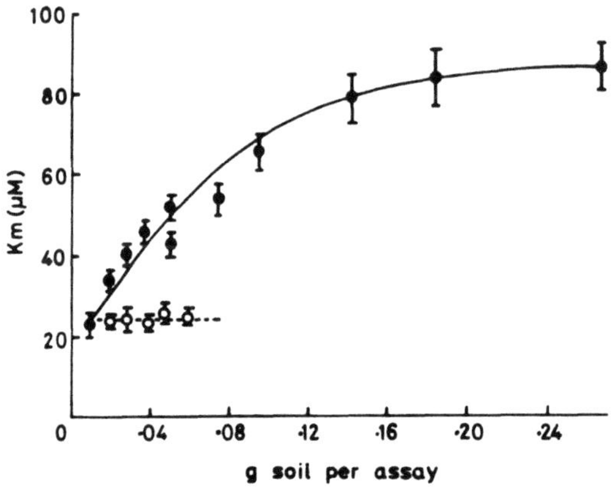

Fig. 5. The variation in Km with the weight of soil aggregates (of 26.5 μm mean sphere radius) used in the assay. The *open circles* denote aggregates which were maintained in suspension throughout the assay, while the *closed circles* denote aggregates which were allowed to sediment out to form a layer through which the substrate had to diffuse.

aggregates of < 64 μm were used in the assay so as to give layers of varying depth through which substrate had to diffuse, a similar pattern emerged (Fig. 5).

The macroscopic Michaelis-Menten model has no mechanisms which can account for the changes observed in the kinetic constants that result from changes to the physical characteristics of the system. To account for these changes, the system was modelled by using a transient state microscopic model, similar to that of Engasser and Horvath (1973), which considered the diffusion of substrate into (and product out of) each soil aggregate and linked the rate of enzyme hydrolysis to the local substrate concentration at each point in that aggregate using the Michaelis-Menten equation (Darrah and Harris, in preparation). The model was shown to give an extremely good fit to the experimental data for the experiment involving different aggregate sizes, which was the only case they modelled.

The model also predicted that changes in the enzyme content of the soil or changes in the substrate diffusion coefficient due to changes in temperature or moisture content would also change the rate of hydrolysis at a particular substrate concentration and hence change the value of Km estimated. Therefore, this analysis would suggest that while the macroscopic Michaelis-Menten model is capable of fitting data from a particular experiment extremely well, its capacity to predict the results from experiments with different physical characteristics is extremely limited. If so, the macroscopic Michaelis-Menten interpretation of soil enzyme behaviour may best be described as an empirical rather than a mechanistic model.

8 Micro-Scale Diffusion of Soil Acidity in Soil

The transformation of ammonium to nitrate via nitrification is one of the most intensively studied microbial processes in soil, due in part, to its key role in controlling the leaching potential of N from the soil to groundwater. Many simulation models of nitrification have been developed incorporating varying levels of mechanistic detail (Frissel and van Veen 1981). The most detailed use the Monod equation to describe the growth of the nitrifiers and a yield expression Y to relate this growth to nitrate production i.e.

$$\frac{dm}{dt} = \mu_{max} \frac{S}{S+Km} \cdot m; \tag{10}$$

$$-\frac{ds}{dt} = \frac{dN}{dt} = \frac{1}{Y}\frac{dm}{dt}, \tag{11}$$

where m is the number (or biomass) of nitrifying organisms per unit volume of soil, μ_{max} is the maximum rate of growth, Km is an affinity constant Y is the growth yield, and S and N are the concentrations of ammonium and nitrate (or nitrite) respectively. This approach is macroscopic is scale as both m and S are defined on a macroscopic or bulk scale. Hence, S is the ammonium concentration in the bulk soil solution.

The first stage of nitrification involves the oxidation of ammonium to nitrite and also produces acid according to the equation:

$$NH_4^+ + 3/2\ O_2 \rightarrow NO_2^- + H_2O + 2H^+.$$

Nitrifiers are sensitive to low pH and in modelling the effects of prolonged incubations it is necessary to take account of the self-inhibition of the process by acid production (Darrah et al. 1986).

Following the addition of ammonium to the soil, the Monod model predicts an exponential increase in the rate of nitrate production with time, and this has been confirmed by experiment (Darrah et al. 1986). However, other workers in similar experiments have shown the rate to be linear with time (Addiscott 1979), a phenomenon that the macroscopic Monod model cannot simulate. In an attempt to resolve these conflicting results, Darrah et al. (1987) developed a microscopic model of soil nitrification. This model assumed that the nitrifying organisms were distributed as clusters or colonies in the soil, formed by repeated cell division, and examined the effects of this microscopic concentration of organisms on the local soil environment surrounding the clusters. One effect of clustering would be that acid production would be confined to a smaller volume of soil than if the organisms were distributed uniformly through the soil and hence the pH of the soil would be more severely reduced in localized regions of the soil. The model simulated the production of acid by nitrifying clusters, the diffusion of this acid away from the cluster, and the inhibition of nitrification arising from the localized pH change. The microscopic model predicted that very steep pH gradients would arise around nitrifying clusters (Fig. 6) and that this localized pH change could give rise to the various patterns of nitrate production found in the literature. Thus, depending on the size of the cluster and on the initial pH of the soil, it was shown that the rate of nitrate production could increase exponentially, remain linear or even decrease with time (Fig. 7). The microscopic model was therefore capable of predicting a wide range of patterns of nitrate behaviour. Although no direct evidence could be produced to support the model, it has been shown that nitrifying clusters exist in the soil on the micro-scale (Molina 1985).

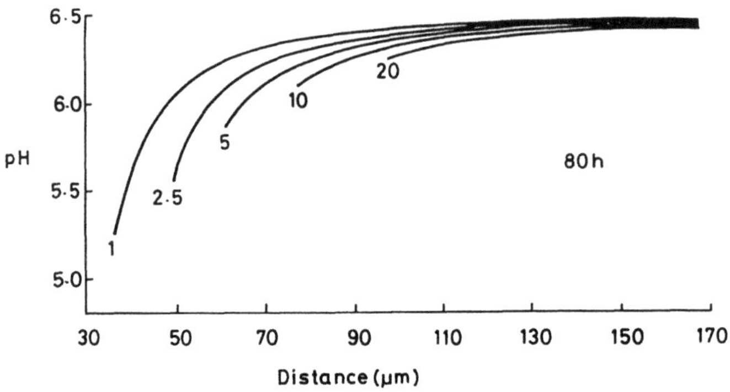

Fig. 6. The predicted gradients of pH in the soil surrounding a nitrifier cluster as a function of the radial position within a soil sphere of 170 μm radius. The *figures* on the curve denote the % volume of the soil sphere assumed to be occupied by the cluster. No pH gradients were assumed to occur within the clusters

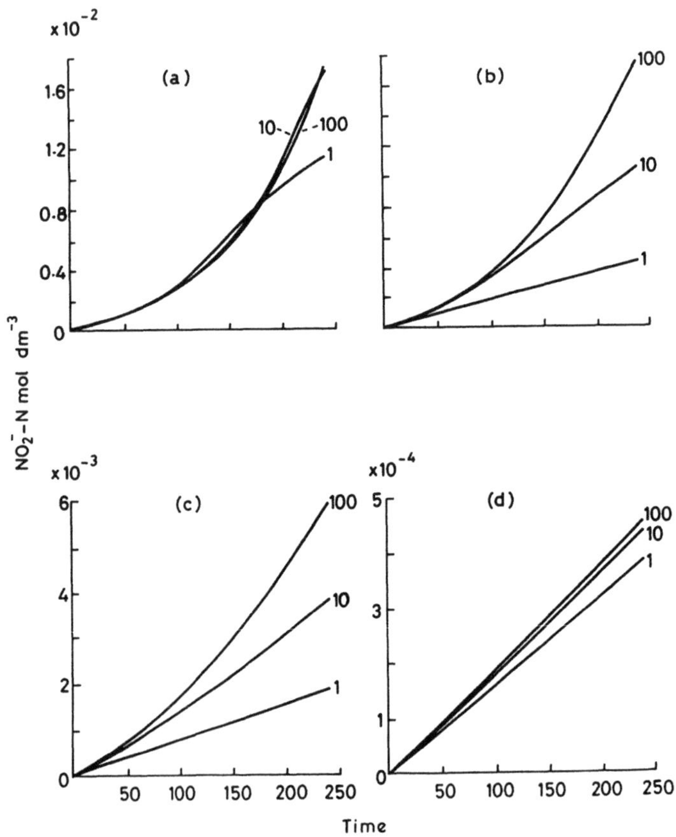

Fig. 7. The different patterns of nitrite production predicted by the model as a function of the % volume of soil occupied by a nitrifier cluster *(figures on curves)* and of the initial soil pH. The initial pHs were **a** 7.5; **b** 6.5; **c** 5.5; **d** 4.5

9 Conclusions

The evidence presented shows that the interpretation of some fundamental biological processes in soil has to take account of mechanisms which may operate on the micro- as well as the macro-scale. Although only micro-scale diffusion limitation has been discussed, other physical and chemical processes, e.g. mass flow and adsorption, may also limit biological activity on the micro-scale. There is also considerable circumstantial evidence to support the view that such micro-scale considerations are important in a much wider soil microbiological context. For instance, most researchers are aware of the difficulties encountered in replicating biological experiments and this may be partially explicable in terms of subtle changes in the physical and chemical characteristics of the soil which may influence the micro-scale transport of microbial substrates. Similarly, the enormous variety of macroscopic equations and models used to simulate the activity and growth of micro-organisms in soil (Bazin et al. 1976; Hunt et al. 1985) may partly be a consequence of ignoring micro-scale processes. It should be remembered that many of these equations, e.g. the Michaelis-Menten and Monod equations, were developed to describe processes which occur in homogeneous solutions, and they may not be directly applicable to heterogeneous media.

It is also clear that the distribution of enzymes and micro-organisms on the micro-scale is not uniform (Ahmed and Oades 1984; Kanazawa and Filip 1986) and that the existence of micro-sites of more intense activity is likely to be very common. The spatial heterogeneity of insoluble microbial substrates is another example of systematic micro-scale variation and some of the effects of this have been examined (Hope and Burns 1985). It is therefore unlikely that the concepts of micro-scale transport limitations will be confined to the examples cited.

Micro-scale transport phenomena may therefore have serious implications for the design and interpretation of experiments as well as for the modelling of soil biological processes. If the predictive capacity of mechanistic models of soil processes is to be extended to deal with changes in the physical and chemical environments in which the biological activity resides, as well as with changes in the intrinsic biological activity itself, then a combination of microscopic and macroscopic mechanisms may have to be used. At present, no satisfactory model has been developed which integrates two or more spatial scales into a single model. Further research on the spatial distribution of the soil biota and their substrates is also required before microscopic mechanisms can be included in mechanistic models.

References

Addiscott TM (1979) Kinetics and temperature relationships of mineralisation and nitrification in Rothamsted soils with differing histories. J Soil Sci 34:343–353

Addiscott TM, Wagenet RJ (1985) Concepts of solute leaching in soils: a review of modelling approaches. J Soil Sci 36:411–424

Ahmed M, Oades JM (1984) Distribution of organic matter and adenosine triphosphate after fractionation of soils by physical procedures. Soil Biol Biochem 16:465–470
Bazin MJ, Saunders PT, Prosser JI (1976) Models of microbial interactions in the soil. CRC Crit Rev Microbiol 4:463–498
Burrough PA (1983) Multiscale sources of spatial variation in soil. 1: The application of fractal concepts to nested levels of soil variation. J Soil Sci 34:577–597
Darrah PR (1982) A fluorimetric assay for determining the kinetics of soil hydrolases. PhD Thesis, Univ Reading, UK
Darrah PR, Harris PJ (1986) A fluorimetric method for measuring the activity of soil enzymes. Plant Soil 92:81–88
Darrah PR, Nye PH, White RE (1986) Simultaneous nitrification and diffusion in soil V. The effects of pH change, following the addition of ammonium sulphate, on the activity of nitrifiers. J Soil Sci 37:479–484
Darrah PR, White RE, Nye PH (1987) A theoretical consideration of the implications of cell clustering for the prediction of nitrification in soil. Plant Soil 99:387–400
Dietschy KM (1978) Effect of diffusion barriers on solute uptake into biological systems. In: Microenvironments and compartmentation. Academic Press, New York London, pp 401–415
Engasser JM, Horvath C (1973) Effect of internal diffusion in heterogeneous enzyme systems: evaluation of true kinetic parameters and substrate diffusivity. J Theor Biol 42:137–155
Engasser JM, Horvath C (1975) Electrostatic effects on the kinetics of bound enzymes. Biochem J 145:431–435
Fridland V (1976) Pattern of the soil cover. Isr Progr Sci Transl, Kater, Jerusalem
Frissel MJ, van Veen JA (eds) (1981) Simulation of nitrogen behaviour of soil-plant systems. Centre Agric Publ Doc, Wageningen, Neth
Greenwood DJ, Berry G (1962) Aerobic respiration in soil crumbs Nature London 195:161–163
Hope CFA, Burns RG (1985) The barrier-ring plate technique for studying extracellular enzyme diffusion and microbial growth in model soil environments. J Gen Microbiol 131:1237–1243
Hunt HW, Cole CV, Elliott ET (1985) Models for growth of bacteria inoculated into sterilized soil. Soil Sci 139:156–165
Irving GCJ, Cosgrove DJ (1976) The kinetics of soil acid phosphatase. Soil Biol Biochem 8:335–340
Jury WA, Sposito G, White RE (1986) A transfer function model of solute transport though soil 1. Fundamental concepts. Water Resourc Res 22:243–247
Kanazawa S, Filip Z (1986) Distribution of micro-organisms, total biomass and enzyme activities in different particles of brown soil. J Microbial Ecol 12:205–215
Mattiasson B (1983) Immobilised cells and organelles. CRC, Boca Raton, pp 3–25
Molina JAE (1985) Components of rates of ammonium oxidation in soil. Soil Sci Soc Am J 49:603–609
Pedersen H, Furler L, Ventatasubramanian K, Prenosil J, Stuker E (1985) Enzyme adsorption in porous supports: local thermodynamic equilibrium model. Biotech Bioeng 27:961–971
Philip JR (1980) Field heterogeneity: some basic issues. Water Resourc Res 16:443–448
Rachhpal-Singh, Nye PH (1984) Diffusion of urea, ammonium and soil alkalinity from surface applied urea J Soil Sci 35:529–538
Radford PJ, Greenwood DJ (1970) The simulation of gaseous diffusion in soils J Soil Sci 21:305–313
Smith GD (1978) Numerical solution of partial differential equations: finite difference methods. Clarendon, Oxford New York
Smith KA (1980) A model of the extent of anaerobic zones in aggregated soils, and its potential application to estimates of denitrification. J Soil Sci 31:263–277
Tiedje JM, Sextone AJ, Parkin TB, Revsbech NP, Shelton DR (1984) Anaerobic processes in soils and gaseous losses of nitrogen. Plant Soil 76:197–212
Tillotson PM, Nielsen DR (1984) Scale factors in soil science. Soil Sci Soc Am J 48:953–959
Wilding LP, Drees LR (1978) Spatial variability: a pedologists viewpoint. Diversity of soils in the tropics. Am Soc Agron Spec Publ 34:1–12
Zaborsky O (1973) Immobilised enzymes. CRC, Boca Raton

Effect of Gels and Coatings on Clay Mineral Chemical Properties

M. Robert and M. Terce[1]

1 Introduction

The experiments dealing with the phenomena of anion and cation fixation in soils, and the interpretations of resulting data take into account the existence of three main independent components: organic matter, clay minerals, and iron and aluminium oxides and hydroxides.

Organic matter is modelled by chemically extracted fulvic or humic compounds; clay minerals are modelled by reference clays occurring only in deposits; oxides and hydroxides are usually synthesized in the laboratory. It is quite clear that natural soils are far more complex and we only begin, using for example high resolution transmission microscopy, to discover the heterogeneity of the so-called soil clay fraction.

The existence of coatings on the clay surface, which can be organic or of iron and aluminium origin, is thought to be of great importance to soil properties, but this is still under debate. There are only a few references in the literature on this aspect because experiments with clay + Fe Al coatings are always more difficult to perform than with independent components.

In the first part of this review we provide evidence of the presence of Fe and Al coatings using some TEM (transmission electron microscopy) illustrations. The second part will present some data which can contribute to the prevision of pollutant fixation, both anionic or cationic, by complex clay/Fe or Al compound systems.

2 Micromorphological Evidence of the Presence of Coatings on Clays

The literature contains only a few examples of natural coatings, most of which were obtained (Jones and Uehara 1973) with TEM by deposit on grids, a method which can introduce artefacts. We now have techniques (Tessier and Berrier 1979) enabling the clay to be fixed in a state not too far from the natural wet state

[1]Station de Science du Sol, I.N.R.A., 78000 Versailles, France

in soil. Under these conditions, it is possible to study the microorganization of clay particles (Tessier 1984) both for reference or soil clays.

Using such methods the fixation of different Fe and Al compounds on a clay surface could also be studied experimentally (Robert et al. 1983, 1987a). Polycations or low range crystalline compounds (designated here as gels or amorphous compounds) are formed by assemblages of small spheres about 3 nm in diameter (Fig. 1a) (Murphy et al. 1976). In the case of iron, such compounds can be localized as coatings, either on the kaolinite basal surface (Fig. 1b) or on the tactoid or between tactoids (Fig. 1c and d); oxides or hydroxides are more generally separate from the clay surface and will thus react as independent components. Saleh and Jones (1984) have shown that ferrihydrite or Si ferrihydrite can also be fixed on kaolinite at pH < 7.

These authors show that at pH $>$ at which the p.z.c. (point of zero charge) value of kaolinite changes, adsorption does not occur. So the formation of coating implies mainly electrostatic charges between positively charged polymers and negatively charged surfaces. When coatings are formed, they can be relatively stable even during pH change, but long-term weathering will detach them from the surface (Robert et al. 1987b). We will see later that organic matter can be incorporated in the coating and can decrease the p.z.c. value of iron or aluminium.

Aluminium species, when positively charged, have the same affinity as iron for the clay surface. They will be in an external position on kaolinite or illite but also in an interlayer position in smectite.

The formation and properties of intergrade minerals are more firmly documented (Rich 1968). Positively charged Al species are fixed at a pH less than 5 and can form islands in the interlayer space, reducing the negative charge and conferring even a positive charge to the mineral. The presence of interlayer iron has been obtained only at very low pH (Carstea et al. 1970), but it is unstable and is hydrolyzed with proton production. It induces the replacement of interlayer iron by aluminium; so iron has mainly an external position on the clay mineral particle.

Thus, the conclusion is that coatings on clays do exist but they generally consist of polycations, polymers or low range crystalline compounds (Robert et al. 1987b). This formation of coatings has two main consequences on clay properties. It affects the clay microorganization: coated clays form microaggregates ranging from 2 or 3 μm to 50 μm (Fig. 1e and f) (Robert et al. 1983, 1987a). Physical properties and especially water retention and hydraulic conductivity are changed due to porosity modification in relation to the microorganization (increase in K values). External and internal surfaces will also vary due to the presence of coating, thus inducing drastic changes in clay chemical properties.

→

Fig. 1. Illustrations of coatings. (**a**) Fe polycation R = 2.5; (**b**) Fe coating on kaolinite basal surface; (**c**) Fe coating on montmorillonite tactoid; (**d**) Fe coating between montmorillonite tactoids; (**e**) montmorillonite microaggregates (with Al coating); **f** illite microaggregate (with Fe coating) **a,b,c,d,f** = TEM (photographed by Clinard-Jaunet); **e** = SEM (photographed by Berrier)

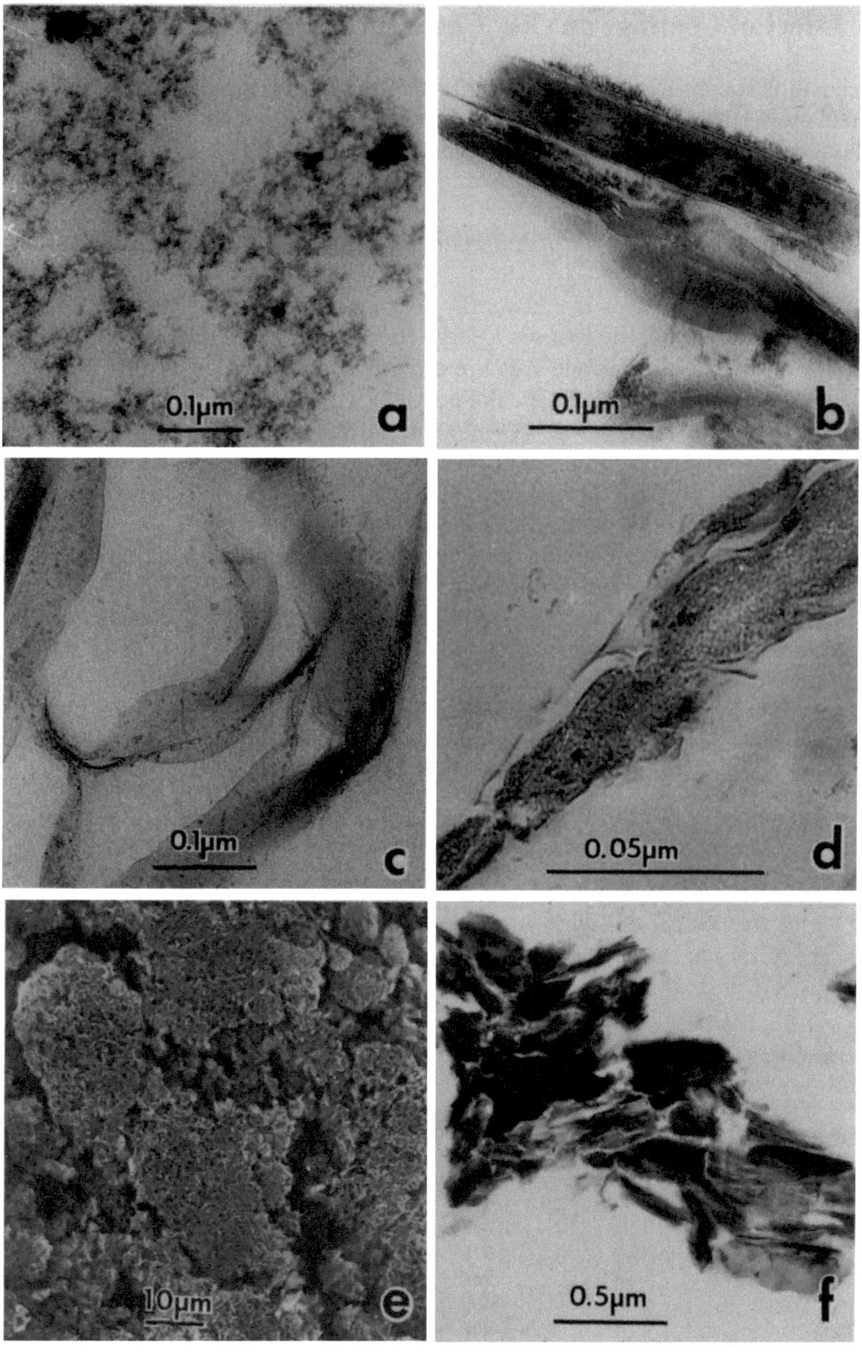

Fig. 1a-f

3 Effect of Coatings on Clay Chemical Properties

We will first consider the surface characteristics of gels and coatings alone and their properties in relation to surface charge and cation fixation. Secondly, the clay-coating association will be examined.

3.1 Surface Characteristics of Amorphous Compounds

Amorphous compounds are generally formed of small particles 2–5 nm and crystallization normally induces a large increase in particle size (× 10). The values of the specific surface area are directly proportional to the particle size. When determining gel surfaces, the ethylene glycol method gives high values ranging from 400 to 800 $m^2 g^{-1}$ (Pyman and Posner 1978) corresponding to both the external surface and internal porosity; results obtained by N_2 fixation seem to correspond to "external surfaces only". Crystalline compounds do not have such an "internal surface" and thus have smaller surface areas.

Table 1 shows surface area results of ferrihydrite obtained with N_2, which range from 279 to more than 400 $m^2 g^{-1}$.

For goethites of different crystallinity (Schwertmann et al. 1985) the surface area ranges from 9 to 153 $m^2 g^{-1}$.

Likewise, to high surface area corresponds a higher surface site due to the OH group. For oxides and hydroxides these groups have been estimated at 4 to 10 nm^{-2}. These charges vary with pH following the reactions which are identical for gels, oxides and hydroxides and broken Al octahedral layers:

$$MOH_2^+ \underset{H^+}{\rightleftharpoons} MOH \underset{OH^-}{\rightleftharpoons} MO^- + H_2O.$$

M^+ and OH^- are called potential determining ions. There must be a particular pH at which the surface charge is zero: this condition is known as the point of zero charge (p.z.c.) or the isoelectric point (i.e.p.). For Parks (1965), the i.e.p. and the

Table 1. Properties of natural ferrihydrites (Schwertmann and Fechter 1982)[a]

Nature of sample	$Fe_{o/d}$	Surface area ($m^2 g^{-1}$)	p.z.c.
Fh (5) (Gt)	0.89	279	7.5
Fh (2)	1.02	433	6.3
Fh (2)	1.02	325	5.3
Fh (2)	0.98	366	5.5
Fh (4) (Gt)	0.87	498	5.4
Fh (2)	1.01	425	6.1

[a] Fh, Ferrihydrite: number in parentheses indicates number of XRD lines; GT, trace of goethite; d, dithionite; o, oxalate.

p.z.c. are identical by definition. According to Hunter (1981) and Lyklema (1983, 1984), the i.e.p. is the pH at which the zeta potential is zero (obtained by any electrokinetic method) and the p.z.c. is the point for which the net surface charge is zero (obtained by potentiometric titration). In the absence of non-specific adsorption of non-potential determining ions, p.z.c. and i.e.p. are identical. Specific adsorption of anions increases p.z.c. and reduces i.e.p. Cations have the reverse effect and the difference between p.z.c. and i.e.p. is a measure of the extent of specific adsorption (Lyklema 1983).

Sposito (1981) gives other definitions of some points of zero charge of solid soil particles, but we prefer the definitions of Hunter (1981) and Lyklema (1984). The results of measured p.z.c. or i.e.p. differ according to authors, probably due to the method of measurement, the sample pretreatments and the presence of surface impurities.

Table 2, which shows the heterogeneity of values, is a compilation of i.e.p. or p.z.c. values found by electrokinetic or electrochemical methods. The i.e.p. or p.z.c. values found in the literature (Table 2) would imply for several compounds the development of only a positive charge surface under natural conditions (pH < 7). But the i.e.p. (or p.z.c.) of mineral constituents can be modified by the presence of silica (Schwertmann and Fechter 1982), organic matter and PO_4

Table 2. Isoelectric points of oxides and hydrous oxides and of some ionic minerals common in soils and sediments (Yariv and Gross 1979)

Precipitate or mineral		pH
$Al(OH)_3$	Amorphous	7.1–9.45
α-$Al(OH)_3$	Crys. gibbsite	5.0
γ-$Al(OH)_3$	Crys. bayerite	6.2–9.25
$AlO(OH)$	Boehmite	9.1–9.4
α-Al_2O_3	Corundum	2.2 (?)–9.2
γ-Al_2O_3		8.0–8.5
$Fe(OH)_2$		12.0
Fe_3O_4	Natural and synthetic Magnetite	6.5
$Fe(OH)_3$	Amorphous	7.1–8.5
α-$FeO(OH)$	Goethite	3.2–6.7
$FeO(OH)$	Limonite	3.6
γ-$FeO(OH)$	Lepidocrocite	5.4–7.4
α-Fe_2O_3	Hematite	2.1 (?)–8.6
γ-Fe_2O_3	Magnemite	6.7
$Mg(OH)_2$		$\approx$ 12
$Mn(OH)_2$		7.0
MnO_2		4.0–4.5
SiO_2	Quartz	2.2
SiO_2	Amorphous, sols or gel	1.8
TiO_2	Natural rutile	4.7
$CaCO_3$	Calcite	9.5
$Ca_5(PO_4)_3(F,OH)$	Fluoroapatite	6.0
$Ca_5(PO_4)_3(OH)$	Hydroxyapatite	7.0

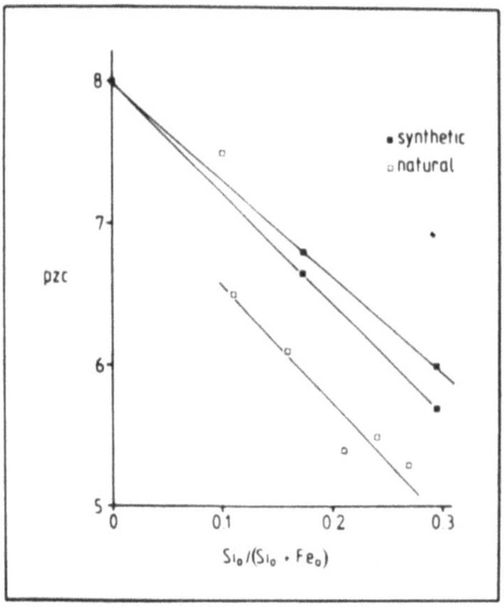

Fig. 2. Point of zero charge of natural and synthetic ferrihydrites in relation to their oxalate-soluble Si/(Si+Fe) mole ration (Schwetmann and Fechter 1982)

(Bowden et al. 1980). For example, a few atoms of Si greatly modify the surface charge of ferrihydrite (Fig. 2).

The presence of these anions both modifies the adsorption properties of Al and Fe compounds in soil and stabilizes their low range crystalline state. When such amorphous compounds are present in coatings on clay surfaces, they represent the interface between solution and clay. They can increase the reactivity of clay surfaces and modify their electrical charge.

3.2 Effect of Coatings on Clay Electrical Charges

Clays have permanent negative charges but also some variable charge on the edge of platelets. It is the general case in 2/1 clays. For 1/1 clays the existence of such a permanent charge is still under debate, but the charge in soil kaolinites seems to stem mostly from isomorphic substitutions (Bolland et al. 1980). It was shown by ESR (Mc Bride 1976) that divalent cations are electrostatically bonded 12 A apart on kaolinite basal surfaces.

Coating the clays can neutralize the permanent negative charge or even reverse the charge to the positive side (Shanmuganathan and Oades 1982, Hendershot and Lavkulich 1983). This is easier to obtain in the order kaolinite > illite > montmorillonite.

The same effect can be obtained experimentally with soils when Fe polycation coating is added (Sumner 1963; Shanmuganathan and Oades 1982). Figure 3, obtained by electrophoresis, shows that the electrophoretic mobility of soil can be reversed, indicating a reversed clay charge.

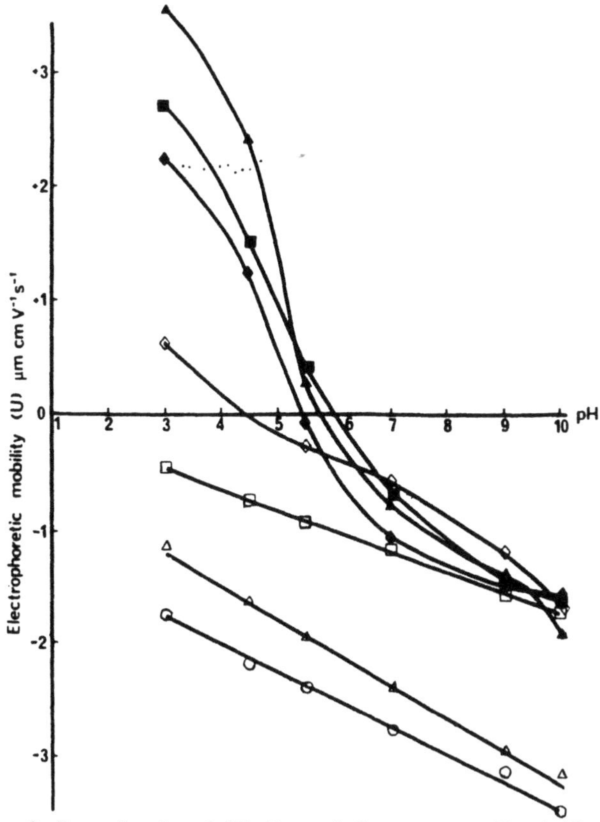

Fig. 3. Electrophoretic mobilities of soils as a function of pH before and after treatment with poly [Fe (III)-OH] cations (○) control; (□) 0.04% Fe; (♦) 0.06% Fe; (◆) 0.08% Fe; (▲) 0.16% Fe; (■) 0.32% Fe. (Shanmuganathan and Oades 1982)

If we refer to van Raij and Peech (1972), the electric charge of some particular alfisoils and oxisols at the soil pH conditions, stems essentially from a net positive charge in the B-horizon and a net negative charge in the A-horizon (Fig. 4). The authors correlated these changes in charge with the presence of oxides or coated clay in B and organic matter in A. If these coatings are removed by chemical treatment, the behaviour is changed. These specific surface charge properties are associated with the modification of both physical properties (flocculation-dispersion, swelling) and fixation properties that we will consider in detail below.

3.3 Effect of Gels or Coatings on Clay Adsorption Properties

We stated previously that data on properties of clay coatings are scarce. Thus, we will first consider the adsorption properties of gels as compared to those of crystalline compounds.

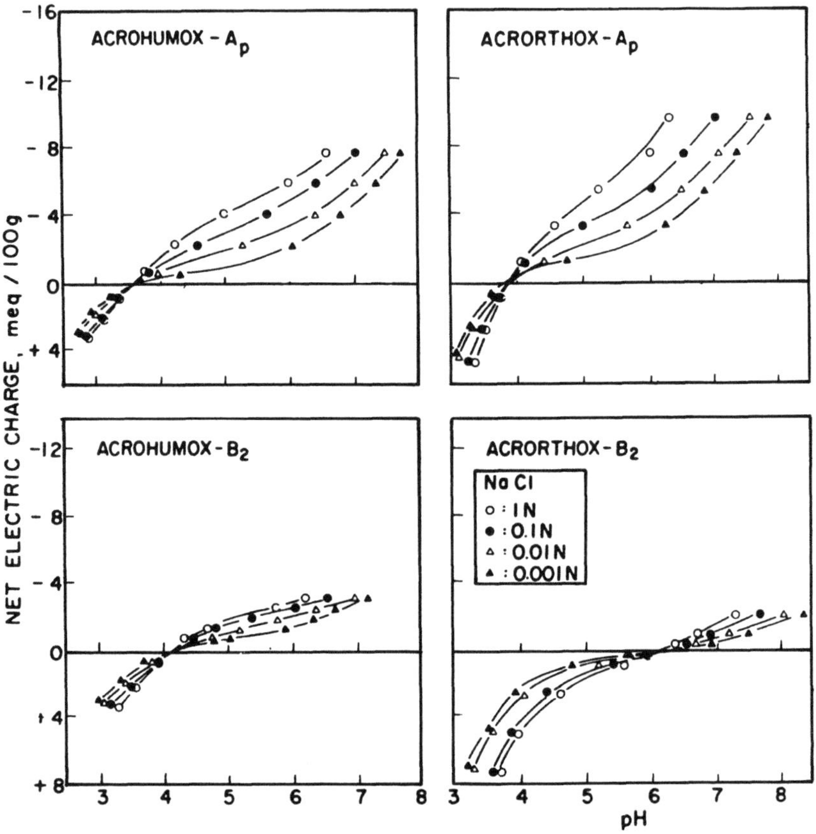

Fig. 4. The net electric charge of soils as determined by potentiometric titration (Van Raij and Peech 1972)

3.3.1 Adsorption of Anions and Cations by Gels

At low pH, gels adsorb organic matter, phosphates, silica and other negatively charged pollutants by the same reactions which govern their retention on oxide and hydroxide (Anderson and Rubin 1981). The mechanisms are complex and still unresolved.

Ions will be attracted or repelled from an oxide surface by electrostatic forces (long-range interactions): in this case, the surface charge given by p.d.i. (H^+ or OH^-) is the main factor of adsorption. But ions will be specifically sorbed (short-range interactions) on the surface (Hingston et al. 1967) as shown experimentally by the shift of i.e.p. or p.z.c. measured by electrokinetic or potentiometric methods (Lyklema 1983). Quantitative models (differing in detail) of both cation and anion adsorption phenomena exist: one uses the surface complexation models (or the surface speciation models), summarized by Anderson

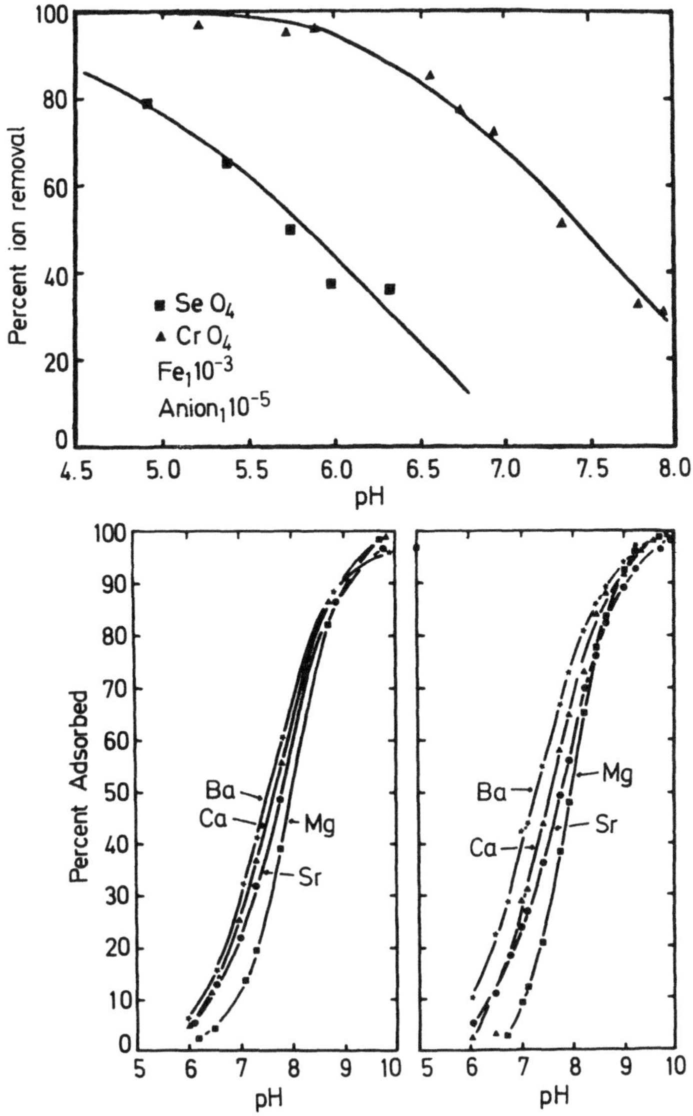

Fig. 5. a Anion adsorption by $Fe_2O_3 \cdot H_2O$ (amorphous) in relation to pH (Benjamin 1983). **b1** Retention of alkaline earth cations by Fe gel (0.093 M Fe) in 1 M $NaNO_3$ from a 10^{-3} M solution of each divalent cation **b2** Four cations together (Kinniburg et al. 1976)

and Rubin (1981), Sposito (1984) and Barrow (1985, 1987). The other uses the basic Gouy Stern theory (Fokkink et al. 1987).

Anion adsorption (Fig. 5a) decreases as pH increases (Benjamin 1983). Conversely, cation adsorption (especially heavy metals) by Al and Fe gels increases with the pH (Kinniburg et al. 1976). Figure 5b shows a symmetrical curve if compared with anion adsorption with an increase of metal adsorption with pH. If the metal retention for the fresh gels and for the corresponding crystalline compounds are compared (Shuman 1977), there is a factor of ten in favour of the gel which stems from the different specific surface area and surface reactivity.

3.3.2 Adsorption of Cations and Other Charged Molecules by Coated Clays

This complex system associates permanent charge, which can have an interlayer (2/1 phyllosilicates) or an external particle position (1/1 phyllosilicates), with a variable charge coating. The latter can be in an interlayer (Al in smectites) or external (the most general case) position (Uehara and Gillman 1980).

With respect to the 2/1 clay minerals and especially the aluminium hydroxide montmorillonite complex, we have up-to-date data (Keizer and Bruggenwert 1987) showing the role of different types of adsorption sites. Figure 6 illustrates the fixation of different pollutant cations: Cu, Zn, Pb and Cd by the Al polymer with three different pH values (pH 5, 6, and 7). The variable charge effect is quite evident with a great increase in element adsorption due to the increase in pH.

The Cu adsorption results obtained by Keiser and Bruggenwert (1987; Fig. 7) permit us to clearly distinguish between:

1. The small effect of clay edges of 2/1 phyllosilicate on adsorption;
2. The slightly greater effect of the clay CEC Cu sorption due to the permanent charge;
3. The more important adsorption on the aluminium hydroxide surface, which is highly pH-dependent;
4. The greatest adsorption which occurs inside the aluminium hydroxide. This fraction is considered as being incorporated into the structure of an interlayered Al hydroxide species. Such a phenomenon occurs for Cu and Zn but not for Pb and Cd (Keizer and Bruggenwert 1987).

Thus, we can see that even for montmorillonite coatings are very effective in increasing cation adsorption. In this example, Al coating was mainly interlayer and it partly neutralized the permanent charge and permitted some specific adsorption.

In the case of iron, the coating will be external to the particle and the montmorillonite will keep its internal, permanent charge and will gain an external variable charge. Even in this case specific adsorption will combine with electrostatic attraction (Fig. 8).

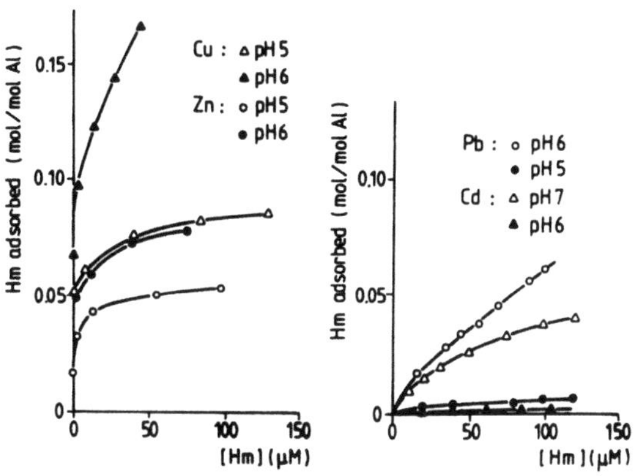

Fig. 6

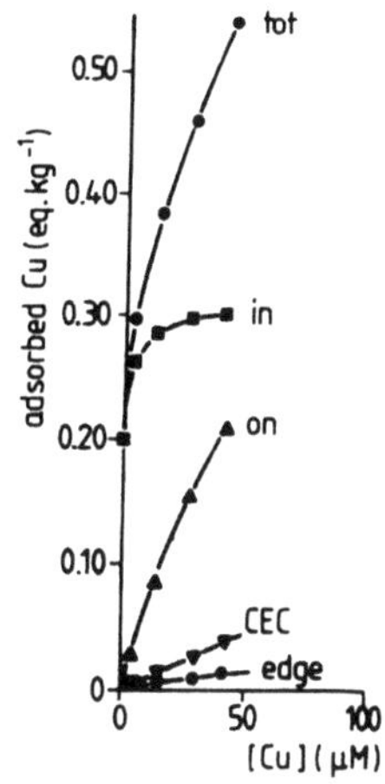

Fig. 7

Fig. 6. Adsorption of Cu, Zn, Pb and Cd on an Al hydroxide coating. (Keizer and Bruggenwert 1987) Hm = heavy metals

Fig. 7. Partition of adsorbed Cu at pH 6 on a montmorillonite with an Al hydroxide coating. *tot* = Total adsorption; *in* = Cu incorporated by Al hydroxide; *on* = Cu adsorbed at the surface of Al hydroxide. *CEC* and *edge* are related to the montmorillonite adsorption (Keizer and Bruggenwert 1987)

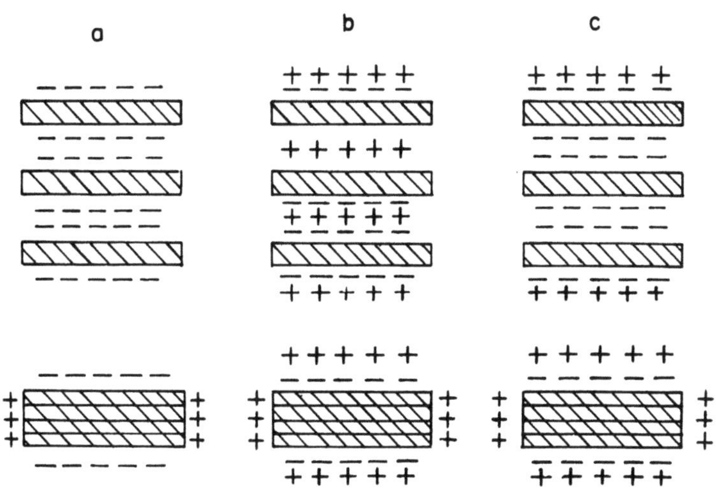

Fig. 8a-c. Schematic representation of (**a**) smectite (*top*) and kaolinite (*bottom*) with (**b**) Al coating (**c**) Fe coating. This pattern is available for pH below p.z.c. Above p.z.c. the positive charge can become negative

There are only a few references concerning coated 1/1 clay mineral and they concern mainly kaolinite behaviour before or after oxide and hydroxide removal (Cavallero and Mc Bride 1984). As stated before, there is ample evidence of the existence of a permanent charge on soil kaolinite (Skjemstad and Koppi 1983). So the case is simpler than for 2/1 clay minerals because coating is always external to the particle and mainly on the basal surfaces (Fig. 8).

This coating will confer a very reactive interface to the kaolinitic mineral, with higher surface and higher OH adsorption sites. These sites can behave in the same way with pH, as the broken OH of the kaolinite edge (Fripiat 1957).

3.3.3 Other Reactions: Fixation of Neutral Molecules or Coprecipitation

We have emphasized the adsorption of charged species, but coating can also interact with neutral molecules. We cite only one example from Calvet and Terce (1975), where polymerized Al adsorbed on montmorillonite or freshly prepared Al polymer can adsorb a pesticide like Atrazine. The neutral molecule is protonated by the following reaction:

$$\begin{matrix} A \ldots\ldots H \\ \quad\quad\quad\;\; \searrow \\ \quad\quad\quad\;\; \nearrow \\ \quad H \end{matrix} O - (AlOH_x)^{(3-x)^+} clay \rightleftharpoons \begin{matrix} AH^+ \\ H \end{matrix} O^- \ldots (AlOH_x)\,(3\text{-}x)^+ \; clay$$

A = neutral Atrazine (hydrogen bond) $\qquad$ AH^+ = protonated Atrazine

Numerous phenomena concerning the reactivity of the Al or Fe gels and coatings are not fully understood as yet. If the first reaction is an electric interaction, chemisorption and coprecipitation can occur giving nucleation of heteropolycationic species at the surface (Mc Bride 1976; Kinniburg and Jackson (1979).

Under certain conditions, Al polymer, external on the interlayer in clay minerals, can be associated with some organic matter (Huang and Violante 1986). Such a mixed polymer will have specific adsorption properties (Levy and Francis 1971).

4 Conclusion

In order to make some predictions about anions and cations or even neutral molecule adsorption in soils, we need a better understanding of some parameters.

One of these parameters concerns the mineral compounds which actually occur in soils. The first applications of HRTEM in soil clay mineralogy show the great variety in species, size and reactivity of soil clays; the latter are therefore very different from the reference clays used in experimental studies. The presence of complex coatings on clay particles must also be better documented because it can change both the physical and chemical properties of clay drastically. In regards to physical properties, the coatings can induce flocculation and particle association and thus completely change the hydraulic properties of soils.

We have shown experimentally that these coatings consist mainly of adsorbed polymers or gels and that this adsorption is the main factor of their stabilization (Robert et al. 1987b).

If the mechanism governing chemical properties of Al and Fe gels and coatings are the same as for crystalline compounds, surface and reactivity are far greater.

Coating has several important consequences regarding ion fixation on clay mineral surfaces. The permanent negative charge, which seems to be the more general characteristic for clay minerals, will be substituted for a charge varying with pH, permitting the fixation of anion adsorption conditions.

The pH value which induces a drastic change in surface charge and adsorption properties (p.z.c. value) shifts considerably particularly due to the presence of other specifically adsorbed ions. So, soil chemical behaviour can be very complex and can change with hydric or biological seasonal behaviour, e.g. the decrease of pH following nitrification inducing aluminium release.

The role of electrostatic phenomena has been privileged because they will govern the direction of the reactions and the first step of the interactions with the clay surfaces, but it is obvious that other surface reactions can develop (Sposito and Schindler 1986). So, an important consequence of the presence of coatings is that they confer specific anion and cation adsorption properties to the soil clay minerals.

Acknowledgements. The authors thank E. Ron (Versailles) for her help in translating and Dr. B. Bar-Yosef for revising the paper.

References

Anderson MA, Rubin AJ Eds (1981) Adsorption of inorganics at solid-liquid interfaces. Ann Arbor Sci Publ, Collingwood Michigan 357 pp

Barrow NJ (1985) Reaction of anions and cations with variable charge soils. Adv Agron 38:183–230

Barrow NJ (1987) Reactions with variable charge soils. In: Revue in plant and soil science. Martinus Nighoff Publ, Dordrecht, 191 pp

Benjamin MM (1983) Adsorption and surface precipitation of metals on amorphous iron oxyhydroxide. Environ Sci Technol 17:686–692

Bolland MDA, Posner AM, Quirk JP (1980) pH Independent and pH dependent surface charges on kaolinite. Clays Clay Min 28:412–418

Bowden JW, Nagapajah S, Barrow NJ, Posner AM, Quirk JP (1980) Describing the adsorption of phosphate, citrate and selenite on a variable-charge mineral surface. Aust J Soil Res 18:49–60

Calvet R, Terce M (1975) Adsorption de l'atrazine par des montmorillonites Al. Ann Agron 26 (6):693–707

Carstea DD, Harward ME, Knox EG (1970) Comparison of iron and aluminum hydroxy interlayers in montmorillonite and vermiculite I formation. Soil Sci Soc Am Proc 34:517–521

Cavallero N, Mc Bride MB (1984) Effect of selective dissolution on pH dependent charge in acid soil clays. Clays Clay Min 32:283–290

Fokkink LGJ, De Keizer A, Lyklema J (1987) Specific ion adsorption on oxides: surface charge adjustment and position stoichiometry. J Colloid Interface Sci 118/2:454–462
Fripiat JJ (1957) Propriétés de surface des alumino-silicates. Bull Gr Fr Argile 11:23–47
Hendershot WH, Lavkulich LM (1983) Effect of sesquioxide coatings on surface charge of standart mineral and soil samples. Soil Sci Soc Am J 47:1252–1260
Hingston FJ, Atkinson RJ, Posner AM, Quirk JP (1967) Specific adsorption of anions. Nature London 215:1459–1461
Huang PM, Violante A (1986) Influence of organic acids on crystallization and surface properties of precipitation products of aluminium. In: Huang PM, Schnitzer M (eds) Interactions of soil minerals with natural organic and microbes. SSSA Spec Publ 17
Hunter RJ (1981) Zeta potential in colloid science. Principles and applications. Academie Press, New York London, 386 pp
Jones RC, Uehara G (1973) Amorphous coatings on mineral surfaces. Soil Sci Soc Am Proc 37:792–798
Keizer P, Bruggenwert MGM (1987) Adsorption of heavy metals by clay-aluminum hydroxide-complexes. Coll OTAN, Ghent, 1987
Kinniburg DG, Jackson ML (1979) Cation adsorption by hydrous metal oxides and clay. In: Anderson MA, Rubin AJ (eds) (1981) Adsorption of inorganics at solid-liquid interfaces. Ann Arbor Sci Publ, Collingwood, Michigan
Kinniburg DG, Jackson ML, Syers JK (1976) Adsorption of alkaline earth, transition and heavy metal cations by hydrous oxide gels of iron and aluminium. Soil Sci Soc Am J 40:796–799
Kuo S, Mc Neal BL (1984) Effects of pH and phosphate on cadmium sorption by a hydrous ferric oxide. Soil Sci Soc Am J 48:1040–1044
Levy R, Francis CW (1971) Adsorption and desorption of cadmium by synthetic and natural organo-clay complexes. Geoderma 15:361–370
Lyklema J (1983) Adsorption of small ions. In Parfitt GD, Rochester CH (eds) Adsorption from solution at the solid/liquid interface. Academic Press, New York London, pp 223–246
Lyklema J (1984) Points of zero charge in the presence of specific adsorption. J Colloid Interface Sci 99/1:100–117
Mc Bride M (1976) Origin and position of exchange sites in kaolinite: Am ESR study. Clays Clay Min 24:88–92
Murphy PJ, Posner AM, Quirk JP (1976) Characterization of partially neutralized ferric nitrate solutions. J Colloid Interface Sci 56:270–283
Parks GA (1965) The isoelectric points of solid oxides, solid hydroxides, and aqueous hydroxy-complex systems. Chem Rev 65:177–198
Pyman MAF, Posner AM (1978) The surface areas of amorphous mixed oxides and their relation to potentiometric titration. 1978. J Colloid Interface Sci 66:85–94
Rich C (1968) Hydroxy interlayers in expansible layer silicates. Clays Clay Min 16:15–30
Robert M, Veneau G, Hervio M (1983) Influence des polycations du Fer et de l'aluminium sur les propriétés des argiles. Sci Sol 3-4:235–251
Robert M, Veneau G, Abreu MM (1987a) Etudes microscopiques d'associations Aluminium-Argiles ou Fer-Argiles. Congr Int Micromorphologie des sols, Paris, 1985, pp 467–474
Robert M, Veneau G, Jaunet AM, Hervio M (1987b) Evolution of Fe and Al polycations in oxides hydroxides in the presence of clay minerals. 6th Meet Eur Clay Groups, Sevilla
Saleh AM, Jones AA (1984) The crystallinity and surface characteristics of synthetic ferrihydrite and its relationship to kaolinite surfaces. Clay Min 19:745–755
Schwertmann U, Fechter H (1982) The point of zero charge of natural and synthetic ferrihydrites and its relation to adsorbed silicate. Clay Min 17:471–476
Schwertmann U, Cambier P, Murad E (1985) Properties of goethites of varying crystallinity. Clays Clay Min 33:369–378
Shanmuganathan RT, Oades JM (1982) Modifications of soil physical properties by manipulating the net surface charge on colloids through addition of Fe(III) polycations J Soil Sci 33:451–465
Shuman LM (1977) Adsorption of Zn by Fe and Al hydrous oxides as influenced by aging and pH. Sci Soc Am J 41:703–706

Skjemstad JD, Koppi AJ (1983) Chemical and charge characteristics of kaolinitic soils of south-east Queensland. Aust J Soil Res 21:271-283
Sposito G (1981) The operational definition of the zero point of charge in soil. Soil Sci Soc Am J 45:292-297
Sposito G (1984) The surface chemistry of soils. Oxford Univ Press, New York; Clarendon Press, Oxford
Sposito G, Schindler PW (1986) Reactions at the soil colloid-soil solution interface. 13th ISSS Congr, Hamburg, Germany, VI, pp 683-699
Sumner ME (1963) Effect of iron oxides on positive and negative charges in clays and soils. Clay Min Bull 29:218-226
Tessier D (1984) Etude expérimentale de l'organisation des matériaux argileux: hydratation, gonflement et structuration au cours de la dessiccation et de la réhumectation. Thèse Doct. d'Etat, Paris, INRA 361 pp
Tessier D, Berrier J (1979) Utilisation de la microscopie électronique à balayage dans l'étude des sols. Sci Sol 1:67-82
Uehara G, Gillman GP (1980) Charge characteristics of soils with variable and permanent charge minerals. I. Theory. Soil Sci Soc Am J 44:250-252
Van Raij B, Peech H (1972) Electrochemical properties of some oxisols and alfisols of the trophics. Soil Sci Soc Am Proc 36:587-593
Yariv S, Cross H (1979) Geochemistry of colloid systems for earth scientists. Springer, Berlin Heidelberg New York, 450 pp

Part B Behaviour of Specific Elements

Introductory Comments. Inorganic pollutants constitute a hazard to the environment due to leaching to underground drinking water, accumulation in edible crops and toxification of microbial communities in soils. Besides general theories, discussed in Part A, specific retention characteristics of pollutants in different soils must be known in order to predict potential contamination hazards under various environmental conditions.

The specific behaviour of fluoride, a potent contaminant unless properly handled at disposal, is discussed by Bar-Yosef, who on the basis of soil and plant-related parameters, suggested criteria for permitted loading of different soils with this element. Gambrell and Patrick showed in their contribution the importance of soil redox potential in regulating the mobility and plant availability of Zn and Cd, two key contaminants in waste materials that may be applied to soils.

The effect of soil solution heavy metal toxicants on microbial activity, and the resulting impact on mineralization processes and microbial degradation of toxic organic chemicals, were reviewed by Domsch. The text and examples provided underscore the importance of this effect, sometimes neglected when considering the adverse effects of heavy metals on the environment.

B. Bar-Yosef

Fluoride Sorption and Mobility in Reactive Porous Media

B. Bar-Yosef, I. Afik, and R. Rosenberg[1]

1 Introduction

Fluorine is the chemically most reactive of all the elements and combines directly at ordinary or elevated temperatures with all the elements other than nitrogen, oxygen and the lighter noble gases. It is the twelfth most abundant element in igneous rocks and ranks third among the most abundant anions. It is not surprising, therefore, that compounds of F should be present in vegetation, water, air and the flesh and organs of all living things.

Fluorine in minor amounts is beneficial to animals and human beings but when absorbed in excessive amounts is toxic. A fluorine concentration of 30 to 50 mg F kg^{-1} in the diet of mammals is considered toxic, whereas 6 to 20 mg kg^{-1} is considered beneficial (Hansen et al. 1958; Underwood 1971). Excessive F content in mammals may cause bone cancer (Bohn et al. 1979, p. 287), chromosome aberration and unscheduled DNA synthesis (Tsutsui et al. 1984). Deficient F levels in mammals result in a high incidence of mottled teeth and caries (Underwood 1971). Plants vary in their response to F. Wheat, navel orange, cabbage, peach, tomato, buckwheat and alfalfa were reported to have reduced yields when the F content in their leaves surpassed a given value specific to each species (Leone et al. 1948; Hansen et al. 1958; Singh et al. 1979). This was suggested to arise from a decrease in cellular Ca and Mg due to complexation or precipitation (Cook et al. 1976), and from the effect of F on enzyme activity (Yang and Miller 1963; Weinstein 1977).

Among the common air pollutants, F is ranked fifth with respect to the amount of plant damage produced in the United States. Application of soil amendments (phosphogypsum), fertilizers (superphosphate) and insecticides (sodium fluoride) create other sources of F pollution.

Relative to the hazard of soil contamination by fluorine, little is known about the partitioning of F between the liquid and solid phases of reactive porous media and about F transport in soils. Solubility diagrams of F in soils developed by Street and Elwali (1983) and Elrashidi and Lindsay (1985, 1986) indicate that within the pH range of 5.5–7.0, CaF_2 is the solid phase that controls F activity, whereas at higher pH values F activity is determined by fluorphlogopite. Fluoride adsorption by goethite was studied by Hingston et al. (1968), who reported

[1]Agricultural Research Organization, The Volcani Center, Institute of Soils and Water, P.O. Box 6, Bet Dagan 50250, Israel

maximum adsorption at pH near the pK of HF. Parfitt (1978) suggested that F^- adsorption by clay minerals involved ligand exchange with a coordinated OH group. Bower and Hatcher (1967) and Omueti and Jones (1977) showed that F adsorption was accompanied, indeed, by OH^- release from the surface. Bower and Hatcher (1967) found that gibbsite, halloyisite and kaolinite adsorbed more F per gram than goethite, montmorillonite and vermiculite, but the comparison was made under variable pH conditions.

Fluoride adsorption by soils was shown to conform to a Langmuir model (Bower and Hatcher 1967; Fluhler et al. 1982) or a Freundlich equation (Barrow and Shaw 1977; Peek and Volk 1985). Barrow (1985) simulated F sorption by goethite as a function of pH with Bowden's electro-chemical adsorption model (1977).

Fluoride mobility in soils was scantily reported in the literature (Fluhler et al. 1982). Tracy et al. (1984) reported that F may move into the groundwater of calcareous soils irrigated with water of high F concentations.

The objectives of this work were to study F retention by porous media differing in chemical characteristics, to relate F mobility in soils to soil chemical properties and to propose criteria for permitted loading of different soils with F that will minimize its movement down the soil profile.

2 Materials and Methods

2.1 F Adsorption by K-Montmorillonite

The < 2 μm clay fraction of Belle-Fourche, S.D. bentonite was obtained by allowing larger particles to settle out of a suspension and decanting the suspension. The clay was K-saturated by washing it three times with 1 M KCl. It was then washed with distilled water and separated in a high-speed centrifuge until the equilibrium solution EC was < 0.05 dS m^{-1}. The salt-free clays were freeze-dried and stored in a desiccator.

Flouride sorption at several pH values was studied by shaking 0.5 g clay in 10 ml plastic bottles containing 50 ml F in 0.01 M KCl. To obtain a constant pH, the addition of F to the suspension was preceded by shaking the 0.5 g clay with 25 ml KCl 0.01 M and adjusting the pH to the desired value using KOH or HCl, until no change in pH was observed after shaking for 24 h. The 25 ml F solutions in 0.01 M KCl, adjusted to the required pH, were then added to the suspension and the pH was re-adjusted. Unless otherwise stated, shaking time was 24 h, at 24 $\pm$ 2°C. A preliminary study showed that the sorption process was practically completed by that time. After equilibrium was reached, the suspensions were centrifuged, and the supernatant was analyzed for pH, EC, Ca, Mg, K, P and F concentrations.

The sorption of F by the clay as a function of pH (ranging from 4.5 to 9.0) was studied as described above except that a constant total amount of F (Q_f = 9.5 μmol g^{-1}) was added to the suspensions. Unless otherwise stated, each experimental point in this and the following experiments was duplicated. When replicates did not coincide, both are presented.

2.2 F Sorption by Soils

The studied soils and their chemical characteristics are presented in Table 1. The sorption isotherms were obtained by adding 50 ml KCl and 50 ml KF solutions to 2.5 g of air-dried, 0.5-mm sieved soil. The KCl concentration was adjusted to give, in all cases, an initial Cl + F concentration of 0.01 M. The suspensions were shaken horizontally in plastic bottles at 24 °C for 24 h. Each isotherm consisted of four initial F concentrations. Sorbed amounts were defined as F added minus F found in the solution after a given time.

The effect of contact time (t) on F sorption was studied in soil No. 2 at a total electrolyte concentration of 0.01 M and two solution-to-soil ratios (V) – 1 and 80 ml g^{-1}. The mixing times were 0.17, 2, 7, 15 and 60 days. Each point was replicated three times.

The effect of V on F sorption was investigated by equilibrating 1, 2, 4, 8, 80 and 800 ml solution g^{-1} soil. Equilibration times were 5 and 48 h; the ionic strength and F added (Q_F) were 0.01 M and 26.3 μmol F g^{-1} respectively. The suspensions were unstirred in order to maintain uniform conditions in dilute and in thick suspensions, since the latter could not be shaken. Each treatment was replicated three times.

Calcium and Mg were analyzed by atomic absorption, K by flame photometer, P colorimetrically and F by Orion Specific Electrode (for details see Bar-Yosef et al. 1988). The Al concentration in the solution (McLean 1965) was in all cases small (2 ± 1 μM) and therefore omitted from F speciation calculations. Speciation and activity calculation from measured total concentrations was described by Bar-Yosef et al. (1988). Ionic strength was approximated from experimental EC values (Griffin and Jurinak 1973).

The coefficients in the kinetic and adsorption models to be presented were fitted from experimental results by using the SAS (1982) NLIN procedure. Adequacy of the models was evaluated based on an analysis of the residuals. Tests for serial correlations and non-normality of the residual distribution were per-

Table 1. Characteristics of the soils investigated[a]

Soil		Characteristics						
Location	Classification	Hygroscopic water	Clay	Lime	Organic matter	CEC	pH[b]	EC
				g kg^{-1}		mmol kg^{-1}		dS m^{-1}
1 Besor	Quartzipsamment	10	43	34	2.0	42	8.8	0.20
2 Gilat	Haploxeralf	23	130	205	10.7	120	7.9	1.03
3 Eden	Typic camborthid	62	370	390	26.7	270	8.1	1.25
4 Akko	Chromoxerert	95	610	15	13.1	556	7.7	0.62
5 Newe Ilan	Haploxeroll	75	550	55	70.0	550	7.7	1.00
6 Golan	Rhodoxeralf	50	330	0	13.6	266	6.6	0.16

[a] 0–20 cm soil layer.
[b] Saturated paste, except at Besor (1:1, water:soil).

formed. In cases where the tests were not significant ($P > 0.05$) the kinetic and adsorption models were assumed to be statistically acceptable. The linear regression coefficients were fitted according to the SAS (1982) STEP program.

2.3 F Movement in Water-Saturated Soil Columns

Air-dried, 3-mm sieved soils were packed in 4.2-cm diameter, 30-cm-long plexiglass cells. The bottom of each cell consisted of a fine screen of stainless steel. Care was taken to obtain a uniform aggregate distribution and the desired bulk density. Some properties of the soils used are given in Table 1. A constant-head buret at the top of the column provided the driving force for the solution flow. The soils were leached with 3.33 mM KF or KCl solutions. The outflowing solutions were collected with a fraction collector. Every treatment was duplicated and the averaged results are presented.

3 Results

3.1 F Adsorption by K-Montmorillonite

Fluorine adsorption isotherms by K-montmorillonite at three pH levels are presented in Fig. 1. The symbols represent experimental results, whereas the solid lines were calculated according to Eq. (1), using the best-fitted adsorption parameters (Table 2).

$$A_F = T[K_F(F) + K_{HF}(HF)]/[1 + K_F(F) + K_{HF}(HF) + K_{OH}(OH)]. \quad (1)$$

Here A(mol g^{-1}) is F adsorption, T is maximum adsorption sites (mol g^{-1}), shared by F^-, HF^o and OH^-; K_F, K_{HF} and K_{OH} are the adsorption coefficients (1 mol^{-1}), of the respective species and ( ) denotes activity in solution. The inclu-

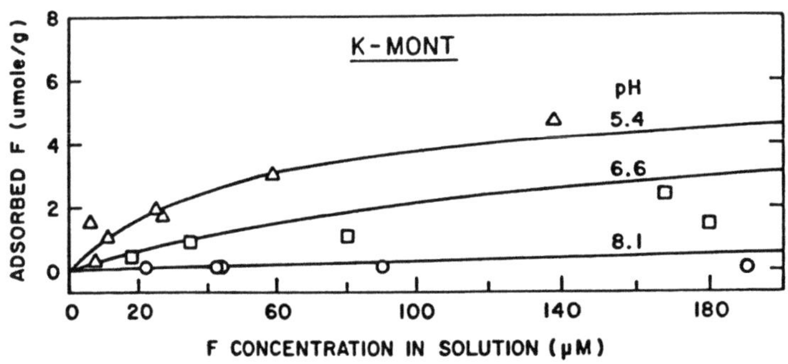

Fig. 1. F adsorption isotherms by K-montmorillonite at three pH values. *Lines* were calculated according to Eq. (1) and parameters given in Table 2

Table 2. Adsorption coefficients [Eq. (1)] for K-montmorillonite

T^a (μmol g^{-1})	K_F^a	K_{HF} (l mol^{-1})	K_{OH}	SEE^b (μmol g^{-1})
5.7**	9.145×10^3*	2.263×10^6*	1.221×10^7**	0.3

[a] **, * t-test significant at $p < 0.01$ and $p < 0.05$ levels.
[b] SEE = standard error of the model estimate.

sion of CaF^+ in Eq. (1) had no practical effect on the results and was therefore omitted. When plotting predicted adsorption lines (Fig. 1), total F concentration in solution, rather than activity, was used, to allow convenient comparison between experimental and predicted results.

The agreement between calculated lines and experimental results indicates that Eq. (1) described F adsorption by the clay as a function of F solution concentration (C_F) at the selected pH values. This supports earlier findings which showed that the competitive Langmuir equation described satisfactorily B and Se adsorption by montmorillonite (Keren et al. 1981; and Bar-Yosef and Meek 1987 respectively), and F adsorption by kaolinite (Bar-Yosef et al. 1988).

To investigate further the effect of pH on F partitioning between solid and liquid phases of K montmorillonite, the "equilibrium" F concentration in solution (C_F) was determined as a function of suspension pH at a given, fixed amount of F in the system (Q_F) (Fig. 2).

It can be seen that C_F increased as the pH was raised from 5.0 to 7.5. At pH 8.5, adsorption was negligible and all added F was found in the solution. The line in Fig. 2 was calculated according to Eq. (1) by using the adsorption

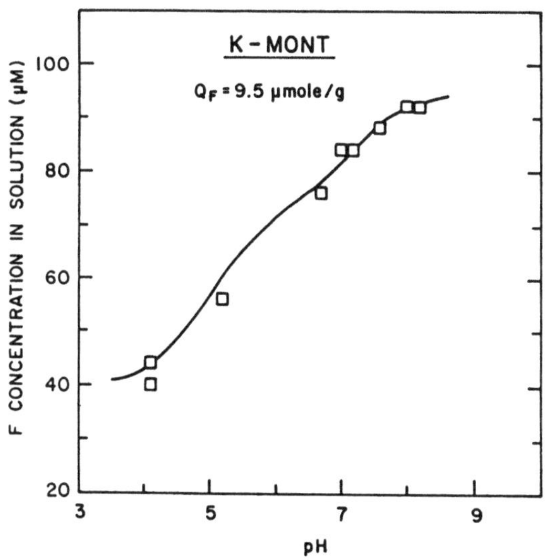

Fig. 2. F concentration in solution of K-montmorillonite suspension as a function of suspension pH at a constant level of F in the system (Q_F = 9.5 μmol g^{-1} clay). *Line* was calculated according to Eq. (1) and parameters given in Table 2

coefficients in Table 1. The compatibility between data and computed results is satisfactory.

To distinguish between adsorption and precipitation reactions, the activities of F^- and Ca^{2+} at each experimental point in Figs. 1 and 2 were compared with the CaF_2 ion-activity product. In all cases, the experimental points were undersaturated, indicating that under the studied conditions CaF_2 was not the solid phase controlling F activity in solution.

3.2 F Sorption by Soils

The rate of F sorption by a test soil (Gilat) is presented in Fig. 3. In an unstirred suspension of 80 ml g^{-1} soil (V = 80) and Q_F of 26.3 or 105.3 μmol F g^{-1} soil, slight variations in C_F occurred between 24 h and 60 days of contact. Similar results were obtained by Peek and Volk (1985). In denser suspensions (V = 1), C_F decreased with time for approximately 16 days and then stabilized at a concentration somewhat higher than at V = 80. The sorption kinetics at V = 1, Q_F = 26.3 and V = 80, Q_F = 105.3 could be approximated by the Elovich equation (inset, Fig. 3). The data indicate that at V = 80, 24 h is a suitable time for F sorption studies.

Fluorine sorption isotherms by six different soils (Table 1) are presented in Fig. 4. Numbers near lines are the "equilibrium" pH values of the soils. The range for each isotherm stemmed from OH^- release by the added F (Parfitt 1978). The

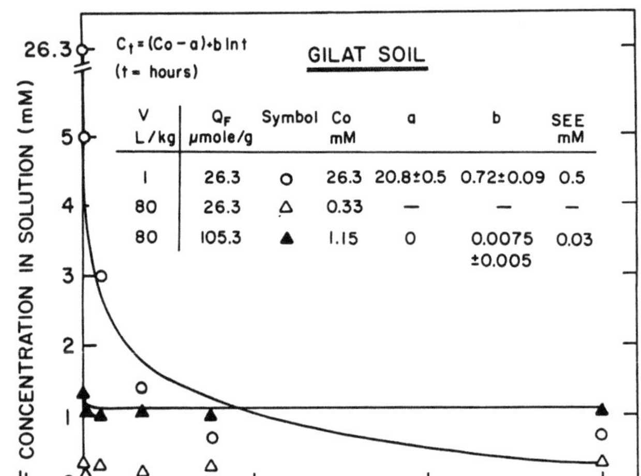

Fig. 3. Fluoride concentration in solution as a function of time of two solution-to-soil ratios (V) and two levels of F (Q_F). Co and SEE are the initial F concentration in solution and the SE of the regression estimate respectively. The ± gives SE of the parameter estimate

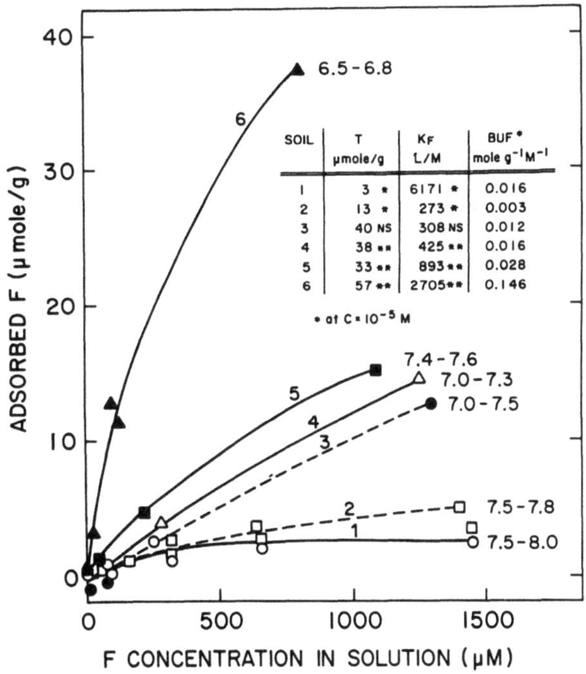

Fig. 4. Fluoride adsorption isotherm by six soils. *Symbols* are experimental results; *Lines* were calculated according to Eq. (1) by using the inset parameters (K_{HF} = O). •, •• t-test significant at the $P < 0.10$ or $P < 0.05$ level. NS = not significantly different than 0. *Numbers near lines* give the suspension pH range. *Numbers above lines* refer to Table 1. *BUF (inset)* is the F buffer capacity of the soils. Electrolyte concentration 0.01 M; equilibration time = 24 h

lines in Fig. 4 were calculated according to Eq. (1) by using the best-fit coefficients presented in the inset, and assuming that K_{OH} = O. The competition of OH^- on adsorption sites was omitted, since at each soil the sorption isotherm was studied only at the natural pH of the soil. In such a case, the effect of soil pH on F adsorption should be expressed through K or T of the soil. The coefficient K_{HF} was estimated by the best-fit procedure to be not significantly different from zero in all studied soils and is therefore not presented.

The studied soils differed in their K_F and T values (inset, Fig. 4). Except in soil 3, these coefficients were significant in all soils. Soil 3 was exceptional due to release of native F into the solution, which masked the adsorption of added F at the low concentration range.

The adsorption constants were correlated by a stepwise linear regression with soil properties. The best single-term model for T was $T = -23.04 \cdot pH + 210.2$ ($R^2 = 0.71$, $P = 0.03$). The best two-term model was: $T = -18.95 \cdot pH + 0.233 \cdot HYG + 165.3$ ($R^2 = 0.84$, $P = 0.07$), HYG being the hygroscopic water content of the soils. This property was shown by Banin and Amiel (1970) to be strongly correlated with soil clay content (CLAY). The best three-term model, which accounted for 96% of the variability in T, was: $T = 16.06 \cdot pH - 0.20 \cdot CEC + 0.223 \cdot CLAY + 140.43$ ($R^2 = 0.96$, $P = 0.05$), where CEC is the soil cation exchange capacity. All units in these, and in following regression equations, are as in Table 1.

The coefficient related to the energy of F adsorption, K_F, was best correlated in a one-term model with soil electrical conductivity (EC): $K_F = -3967 \cdot EC + 4612$ ($R^2 = 0.61$, $P = 0.07$). The best two-term model was $K_F = -4494 \cdot EC + 1706 \cdot pH - 8352$ ($R^2 = 0.87$, $P = 0.05$). No other variables met the 0.25 significance level for entry into the model.

The presented correlations indicate that factors related to clay content in soils (CLAY, HYG, CEC) affect T, but not K_F. Soil pH was highly and inversely correlated with T, and to a lesser extent with K_F. The correlation between T and pH probably stemmed from the effect of pH on the charge density of clay edges. Organic matter and lime content in soils did not correlate significantly with either T or K_F. The negative (significant) correlations between EC and K_F is surprising, since all studied suspensions (40 ml g^{-1}) had an electrolyte concentration of 0.01 M. It is possible that EC reflects some differences among soils which stem from different clay and metal-oxide compositions, and these determine K_F.

The F buffer capacity of soils ($BUF = dA/dC_F = K_F \cdot T/[1 + K_F \cdot C_F]^2$) indicates expected variations in C_F due to incremental changes in adsorbed F (note that for the studied soils K_{OH} and K_{HF} are zero). Elrashidi and Lindsay (1985) reported that the average F concentration in the solution of 109 soils was $\sim 6\,\mu M$. The BUF values of our tested soils at a similar C_F (10 μM) are summarized in the inset of Fig. 4. Even though soil No. 1 had a small T, its BUF exceeded and equalled the BUF of soils 2 and 4 respectively, which had higher T but lower K_F values. A stepwise linear regression on BUF showed that the best one- and two-term models were $BUF = -0.062 \cdot pH + 0.522$ ($R^2 = 0.68$, $P = 0.04$) and $BUF = -0.055 \cdot pH - 0.053 \cdot EC + 0.507$ ($R^2 = 0.88$, $P = 0.04$). The third best correlated variable was the soil lime content (LIME): $BUF = -0.060 \cdot pH - 0.086 \cdot EC + 0.00013 \cdot LIME$ ($R^2 = 0.93$, $P = 0.10$).

The effect of solution-to-soil ratio on C_F after 5 and 48 h of equilibration at a fixed amount of F ($Q_F = 26.3\ \mu mol\ g^{-1}$) is depicted for the Gilat soil in Fig. 5. In the range of 800 to about 6 ml g^{-1}, the data points at both equilibration periods closely follow the theoretical line derived from Eq. (2). This equation assumes that F is either adsorbed according to Eq. (1), or is found in the solution phase:

$$Q_F = A + C_F \cdot V. \tag{2}$$

The model gives at each V a predicted C_F value which is lower than C_F obtained by assuming a mere dilution of F in the soil solution (A = O, dashed line). At $V < 4$ ml g^{-1}, the experimental C_F values are appreciably lower than the calculated ones, the discrepancy increasing with time. To check whether the deviations stemmed from unaccounted for precipitation reactions, F^- activity in equilibrium with fluorite (CaF_2) was calculated from known Ca^{2+} activities and K_{sp} of the mineral (Lindsay 1979). The calculated F^- activity was converted into F concentration (filled symbols) by using the Debye-Huckel equation. Between

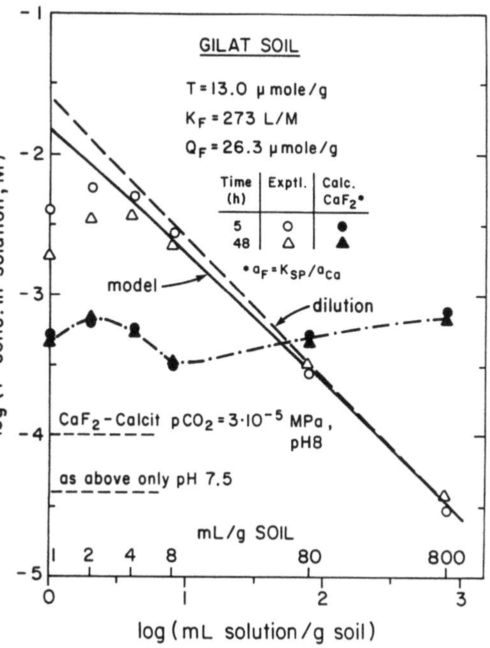

Fig. 5. Effect of solution-to-soil ratio and equilibration time on F solution concentration (C_F) in Gilat soil. *Open symbols are experimental C_F values; filled symbols* are C_F values calculated from K_{sp} of CaF_2 and known Ca activities. The *line* connecting the points was fitted by hand. *Solid line* was calculated according to Eq. (2) by using parameters in Fig. 2. *Broken line* was calculated by assuming a dilution of the soil solution. *Horizontal lines* give C_F for a CaF_2-calcite system at the given pCO_2 and pH

1 and about 50 ml solution g^{-1}, experimental C_F results (open symbols) were supersaturated with respect to CaF_2, indicating that precipitation reactions were occurring at an unknown rate. The possibility that a significant fraction of total Ca in solution was complexed by an organic agent is not expected to change this conclusion. For example, if the fraction of complexed Ca is 0.50, the calculated log F line (filled symbols) is shifted upward by 0.15 units, and is still appreciably below the log F data points. At $V > 50$ ml g^{-1}, C_F was undersaturated with respect to CaF_2, and obeying the adsorption model [Eqs. (1), (2)]. When precipitation and adsorption co-exists, C_F cannot be predicted from Eq. (2), since Q_F is undefined due to the unknown amount of F precipitated.

Fluoride log-activity in a CaF_2-calcite system at P_{CO2} of 3×10^{-5} MPa and pH of 8 and 7.5 is -4.0 and -4.4 respectively (Elrashidi and Lindsay 1985). These pH values resemble the pH in the Gilat suspensions in the adsorption isotherm experiment. Since Gilat soil contains 205 g lime kg^{-1} soil, it is expected that at the above mentioned P_{CO2}, the line connecting the filled symbols in Fig. 5 will descend further with time to approach these log a_F values. It is possible that C_F will be controlled by fluorphlogopite of fluorapatite as well. In a fluorphlogophite calcite/dolomite system, pH = 7.5, the expected log a_F is -5.6 (Elrashidi and Lindsay 1985). However, the data in Fig. 3 show that at V = 80, C_F in the studied soil varied only slightly between 1 and 60 days, indicating that these expected precipitation reactions are very slow.

3.3 F Movement in Water-Saturated Soil Columns

The concentration of F in the inflowing solution (C_{in} = 3.3 mM) was about eight-fold greater than the theoretical C_F in equilibrium with crystalline CaF_2. The breakthrough curves of the Besor, Akko and Gilat soils are presented in Fig. 6, 7 and 8 respectively. The effluent volume at which steady state occurred (outflowing $C_F = C_{in}$) increased with elevating the maximum F adsorption capacity of the soil (Table 3). Observed fluctuations in C_F with effluent volume are attributed to experimental variability. The lines in the figures were hand-drawn for convenience. F retention by the soils, estimated as the difference between inflowing and outflowing F, was compared with predicted F adsorption at equilibrium with C_{in}. Prediction was done according to Eq. (1) and appropriate adsorption parameters (inset, Fig. 4) of the soils. As can be in Table 3, the agreement between experimental and computed result is fair.

Experimental C_{Ca} results at effluent volumes, where $C_F = C_{in}$, can be used to calculate corresponding theoretical CaF_2-controlled F concentrations. Leachate-C_F values thus obtained varied between 150 and 190 μM, compared with C_{in} which was 3.3 mM. This indicates that F concentration in the solution flowing inside the column was supersaturated with respect to CaF_2. One may conclude, therefore, that under the experimental conditions and studied soils, short-term F retention and mobility were governed by adsorption rather than precipitation reactions.

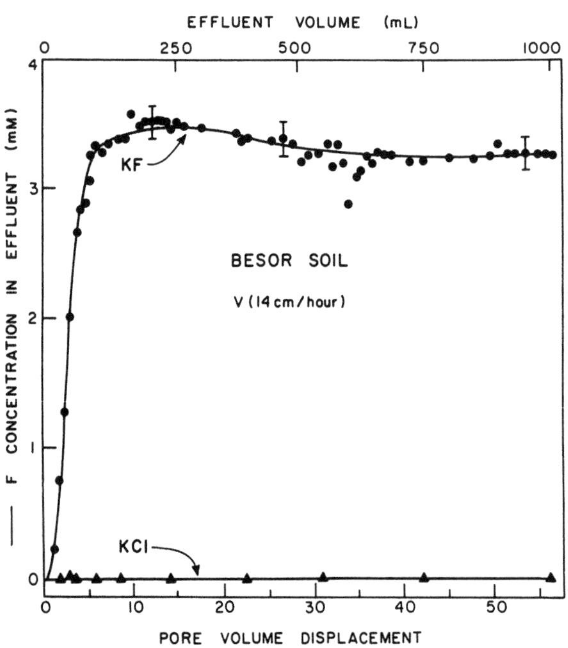

Fig. 6. Effluent F concentration versus effluent volume in water-saturated soil columns leached with either KF or KCl at a concentration of 3.3 mM; besor soil. *Bars* denote maximum deviations from the mean; v denotes solution flow velocity

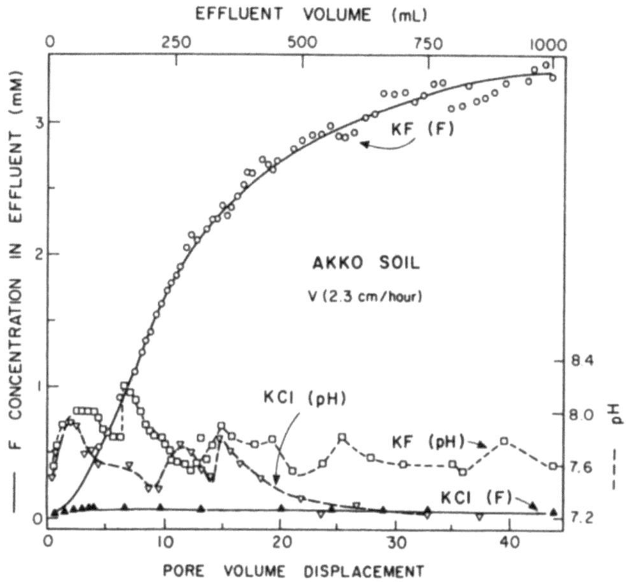

Fig. 7. As in Fig. 6, only for Akko soil

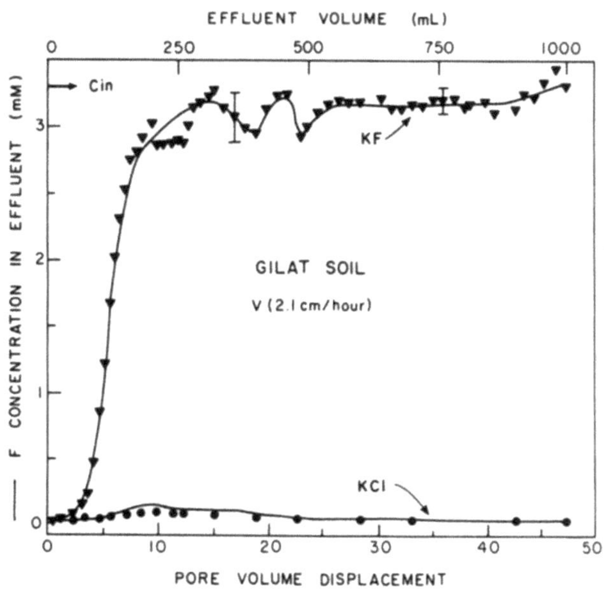

Fig. 8. As in Fig. 6, only for Gilat soil

Table 3. Summary of soil-column experiments with inflowing KF solutions (C_{in} = 3.3 mM l^{-1} KF)

	Soil		
	Besor	Akko	Gilat
PV[a] at which $C_F^a = C_{in}$	7	36	15
Water flow rate (cm h^{-1})	14.0	2.3	2.1
Exptl. F retention[b] (μmol g^{-1})	2.5	18.5	7.3
Predicted[c] F adsorption at $C_F = C_{in}$	2.8	22.2	6.2
C_{Ca} at $C_F = C_{in}$ (mM l^{-1})	1.1	1.1	1.7
Theoretical C_F^d (μM)	190	190	153
pH at $C_F = C_{in}$	7.4	7.5	7.9

[a] PV = pore volume; C_F = effluent F concentration.
[b] F retention = $(V \cdot C_{in} - \int C_F dV/W$; V = total inflowing solution volume (ml); W = soil weight/column (g).
[c] Calculated according to equation (1) and Fig. 4 (inset) parameters for $C_F = C_{in}$.
[d] $\sqrt{K_{sp}(CaF_2)/(C_{Ca} \text{ at } C_F = C_{in})}$.

4 Discussion

In accordance with earlier results (Fluhler et al. 1982), short-term (24 h) F partitioning between solid and solution phases of neutral to calcareous soils was satisfactorily described by a Langmuir adsorption model. By including competition among F^-, OH^- and HF_0 on common adsorption sites [Eq. (1)], F partitioning in K-montmorillonite as a function of pH could be simulated as well. Former models of F sorption by clay minerals (Bower and Hatcher 1967) or soils (Peek and Volk 1983) were unable to predict pH effects on F adsorption.

The time needed to obtain quasi-equilibrium in F sorption reactions was found to be inversely proportional to the suspension's solution-to-soil ratio (V). While at V = 80 ml g^{-1} equilibrium was attained within a few hours, the time needed at V = 1 ml g^{-1} was 10 to 15 days (both suspensions unstirred). The different rates probably stemmed from retarded accessibility of F to adsorption sites at V = 1 relative to V = 80 due to enhanced masking of binding sites on clay edges by the permanent negative electric field of soil-clay faces (Sposito 1984).

The maximum adsorption capacity parameter T was found to be positively correlated with soil-clay content and negatively correlated with CEC. This agrees with the fact that T of K-montmorillonite was estimated to be smaller than T of the studied soils. The mechanism involved may be related to the effect of the overlapping edge-face electric fields mentioned above.

According to Elrashidi and Lindsay (1985), F^- activity in soil solutions ranges between 0.16 and 66 μmol l^{-1}, with an average concentration of ~6 μmol l^{-1}. A similar value was reported by Larsen and Widdowson (1971). At this mean activity and pH between 7–8, the expected F adsorption by montmorillonite and soils is small. Since adsorption-desorption are considered to be faster reactions than precipitation-dissolution (Enfield et al. 1981), the adsorbed amount, even though very small relative to total F in soils, may serve as a readily available pool for plants. This pool is sensitive to variations in pH (Fig. 2), and can be rapidly recharged upon addition of F to the system.

Soils with high F adsorption and buffer capacities are most suitable for F disposal, since they most effectively resist variations in soil solution F concentration. Current data and correlations indicate that such soils have low to intermediate pH, low EC and high content of low-CEC clays.

Fluoride retention by soil columns was reasonably predicted by the Langmuir equation and relevant adsorption coefficients obtained in batch experiments. A similar agreement was reported by Fluhler et al. (1982) in acid soils. In calcareous soils, the same authors found that F retention consisted of only 50% of the predicted sorption, which is appreciably less than the current compatibility.

By comparing Langmuir's adsorption with the soluble-F added to the soil via the water (C_{in} x z, z being the water head of added solution, cm), the depth (h, cm) at which C_F equals C_{in} (mol F l^{-1}) can be estimated:

$$h = 0.001 \cdot z \cdot (1/K_F + C_{in})/(\rho \cdot T). \quad (3)$$

Here, ρ is the soil bulk density, and K_F and T are defined in Eq. (1). Equation (3) is based on the mass balance relationship $C_{in} \cdot z = h \cdot \rho \cdot A \cdot 1000$, where A is Langmuir's adsorption in equilibrium with C_{in}. It is assumed that the amount of F in the soil solution is negligible compared with adsorbed F. When F is added to the soil as an insoluble F-mineral and leached with H_2O, C_{in} can be replaced by the F-mineral, water-soluble F concentration. This relationship might be helpful in estimating F movement down the soil profile under various soil and leaching volume conditions.

Acknowledgement. This study was supported by BARD, the Binational Agricultural Research and Development Fund, Grant No. I-367-81.

References

Banin A, Amiel A (1970) A correlative study of the chemical and physical properties of a group of natural soils of Israel. Geoderma 3:185–198
Bar-Yosef B, Meek D (1987) Selenium sorption by kaolinite and montmorillonite. Soil Sci 144:11–19
Bar-Yosef B, Afik I, Rosenberg R (1988) Fluoride sorption by montmorillonite and kaolinite. Soil Sci 145:194–200
Barrow NJ (1985) Reaction of anions and cations with variable charge soils. Adv Agron 38:183–230
Barrow NJ, Shaw TC (1977) The slow reaction between soil and anions: b. effect of time and temperature of contact on fluoride. Soil Sci 123:265–278
Bohn HL, McNeal BL, O'Connor GA (1979) Soil chemistry. John Wiley & Sons, New York
Bowden JM, Posner AM, Quirk JP (1977) Ionic adsorption by variable charge mineral surfaces. Theoretical charge development and titration curves. Aust J Soil Res 15:121–136
Bower CA, Hatcher JT (1967) Adsorption of fluoride by soils and minerals. Soil Sci 103:151–154
Cook JA, Johnson MS, Davison AW, Bradshaw AD (1976) Fluoride in plants colonizing fluorspar mine waste in the Peak District and Wearsdale. Environ. Pollut 11:9–23

Elrashidi MA, Lindsay WL (1985) Solubility relationships of fluorine minerals in soils. Soil Sci Soc Am J 49:1133–1136
Elrashidia MA, Lindsay WL (1986) Chemical equilibria of fluorine in soils: a theoretical development. Soil Sci 141:274–280
Enfield CJ, Phan T, Walter DM, Ellis R, Jr. (1981) Kinetic model of phosphate transport and transformation in calcareous soils. I. Kinetics of transformation. Soil Sci Soc Am J 45:1059–1063
Fluhler HJ, Polomski, Blaser P (1982) Retention and movement of fluoride in soils. J Environ Qual 11:461–468
Griffin RA, Jurinak JJ (1973) Estimation of activity coefficients from the electrical conductivity of natural aquatic systems and soil extracts. Soil Sci 116:26–30
Hansen ED, Wiebe HH, Thorne W (1958) Air pollution with relation to agronomic crops. VII. Fluoride uptake from soils. Agron J 50:565–568
Hingston FJ, Atkinson LJ, Posner AM, Quirk JP (1968) Specific adsorption of anions on geothite. Trans 9th Int Congr Soil Sci 1:669–677
Keren R, Gast RG, Bar-Yosef B (1981) pH-Dependent boron adsorption by Na-montmorillonite. Soil Sci Soc Am J 45:45–48
Larsen S, Widdowson AE (1971) Soil fluorine. J Soil Sci 22:210–221
Leone IA, Brennan EG, Daines RH, Robbins WR (1948) Some effects of F on peach, tomato and buckwheat when absorbed through roots. Soil Sci 66:259–266
Lindsay WL (1979) Chemical equilibria in soils. Wiley-Interscience, New York
Omueti JAL, Jones RL (1977) Fluoride absorption by Illinois soils. J Soil Sci 28:264–272
Parfitt RL (1978) Anion adsorption by soils and soil minerals. Adv Agron 30:1–50
Peek DC, Volk UV (1985) Fluoride sorption and desorption in soils. Soil Sci Soc Am J 49:583–586
Singh A, Chhabra R, Abrol IP (1979) Effect of fluorine and phosphorus applied to a sodic soil on their availability and on yield and chemical composition of wheat. Soil Sci 128:90–97
Sposito G (1984) The surface chemistry of soils. Oxford Univ Press, New York
Street JJ, Elwali AMO (1983) Fluorite solubility in limed acid sandy soils. Soil Sci Soc Am J 47:483–485
Tracy PW, Robbins CW, Lewis GC (1984) Fluorite precipitation in a calcareous soil irrigated with high fluoride water. Soil Sci Soc Am J 48:1013–1016
Tsutsui T, Suzuki N, Ohmori M (1984) Sodium chloride-induced morphological and neoplastic transformation, chromosome aberrations, sister chromatid exchange and unscheduled DNA synthesis in cultured hamster embryo cell. Cancer Res 44:938–941
Underwood EJ (1971) Trace elements in human and animal nutrition. Academic Press, New York London
Weinstein LH (1977) Fluoride and plant life. J Occup Med 19:49–78
Yang SF, Miller GW (1963) Biochemical studies on the effect of fluoride on higher plants. 3. The effect of fluoride on dark carbon dioxide fixation. Biochem J 88:517–522

Cu, Zn, and Cd Availability in a Sludge-Amended Soil Under Controlled pH and Redox Potential Conditions

R. P. Gambrell and W. H. Patrick, Jr.[1]

1 Introduction

1.1 Background

Disposal of sewage sludge and similar waste by land application has been practiced on a small scale in Europe and in this country for 100 years and longer. Only during the last 15 years has the application of municipal sludge to agricultural soils gained widespread research attention and considerable acceptance as a method of waste disposal and resource utilization.

In the United States, the Federal Water Pollution Control Act of 1972 (PL 92–500) stimulated increased interest in the use of terrestrial ecosystems to achieve water quality standards. Section 201(b) requires that technology development consider advanced waste treatment techniques and include the reclaiming and recycling of water, and the confined disposal of pollutants so they will not migrate or cause water or other environmental pollution. The application of municipal sewage sludge to soil-crop systems may be the most environmentally and economically feasible method of meeting these requirements in many instances. Benefits of land application of municipal sludge include: (1) reducing the pollution load of river and coastal waters that have traditionally received sewage wastes, (2) recycling water and nutrients through crops to recover these valuable resources, (3) more rapid inactivation or destruction of potentially serious microbial pathogenic organisms than some conventional wastewater treatment methods (Braids et al. 1970), and (4) possible improvement of marginal soils for developing land for parks and other urban uses (Peterson et al. 1971).

The most serious limitation of applying municipal sludge to agricultural soils concerns the fate and effects of pathogens and toxic substances that may be associated with municipal wastes. Such materials include bacterial and viral infectious agents, synthetic organic toxins, such as pesticides and PCB's, and toxic heavy metals. Toxic heavy metals are of interest because of the tendency for these metals to accumulate in soils and to be taken up by plants. Crop uptake of toxic metals may increase the dietary intake by livestock and humans who consume these plants as a food source.

---

[1]Laboratory for Wetland Soils and Sediments, Center for Wetland Resources, Louisiana State University, Baton Rouge, LA 70803, USA

Extensive agricultural research has shown that a number of interacting physical, chemical, and biological factors affect the availability of nutrients and toxins in soil-plant systems. For example, the kind and amount of clay minerals and soil organic matter, and the presence of additional chemically reactive or adsorptive soil components, such as sulfides and hydrous oxides, play an important role in regulating metal bioavailability. Plant species grown and management of soils, crops, and waste applications are also important factors. Fortunately, most soils – even those which are highly contaminated – have a large capacity to immobilize toxic metals so that only a small portion of the total metals applied is available to organisms. However, all of the metals applied with sludge applications must be considered potentially available for plant uptake. Because of the probable long-term accumulation of potentially available toxic metals in sludge-amended soils and the dynamic nature of soil-plant systems, it is imperative that we understand the factors and processes affecting metal availability.

Soil pH and oxidation-reduction conditions are two physicochemical parameters that affect metal mobilization and immobilization processes. Considerable research has been conducted on the effects of soil pH on the chemical availability and plant uptake of trace and toxic metals by crop plants grown on soils treated with municipal sludge. These studies have demonstrated that an increase in soil acidity usually enhances plant availability of trace and toxic metals. In contrast, very little is known of the effects of soil oxidation-reduction conditions on toxic metal availability. Soil oxidation-reduction conditions are probably as important as soil pH in regulating the availability of metals. Most soils currently receiving municipal sludge are upland soils classified as moderately well- to well-drained, and such soils are generally considered aerobic or oxidized. However, wetland soils used for rice, or natural wetland habitats, may sometimes be candidates for sludge amendments. In addition to the range in general soil oxidation types encountered, it is likely that the addition of large quantities of sludge (frequently as a water slurry) containing high levels of organic matter may contribute to intermittent, strongly reduced conditions in part or all of the rhizosphere of an upland soil that is normally considered well-drained and oxidized. Thus, a wide range of oxidation-reduction conditions may be encountered in sludge-amended soils possibly affecting the degree of plant availability of toxic metals applied with sludge.

1.2 Literature Review: Soil Redox Effects on Plant Uptake of Metals

Oxidation-reduction conditions affect many of the processes that regulate the chemistry and availability of trace and toxic metals. A list of these regulatory processes would include precipitation as insoluble sulfides under highly reduced conditions (Krauskopf 1972; Morel et al. 1973; Engler and Patrick 1975), formation of discrete carbonates, metal oxides, or hydroxides of low solubility (Morel et al. 1973), or adsorption to colloidal hydrous oxides of Fe and Mn

(Taylor and McKenzie 1966; Jenne 1968; Lee 1973; Windom 1973; Loganathan et al. 1977), primarily in aerobic, neutral or alkaline environments, and complex formation with soluble and insoluble organic matter under all conditions of pH and oxidation intensity (Schnitzer and Skinner 1966; Stevenson and Ardakani 1972; Verloo and Cottenie 1972; Leland et al. 1973).

Weeraratna (1969) reported that Mn uptake by rice from wetland soils was greater under flooded conditions than under unflooded conditions. Conversely, they found Mn uptake from an upland soil was less under flooded conditions compared to unflooded conditions, but this was thought to result from increased Fe availability and subsequent Fe toxicity to the plants grown in the flooded, upland soil. Where Mn availability was increased by flooding, it was attributed to the increased levels of soluble Mn in reduced (flooded) soils as the less soluble manganic (Mn^{4+}) forms were reduced to the more mobile manganous forms (Mn^{2+}).

Clark et al. (1957) also reported that Mn availability to rice was increased by submerged soil conditions.

Mandal (1962) reported on yield and the N, Fe, and Mn content of rice as a function of the oxygen content of the atmosphere overlying the floodwater. The anaerobic and semi-anaerobic treatments resulted in a greater accumulation of both Fe and Mn in the shoot tissue than was found in the semi-aerobic and aerobic treatments. The aerobic treatments were thought to favor stability of the sparingly soluble oxides and hydroxides of these metals, reducing their availability to plants.

Jones and Etherington (1970) reported on the effects of waterlogging on the uptake of Fe and Mn by two marsh species, *Erica cinerea* and *Erica tetralix*, in a study to determine factors favoring the establishment of *E. tetralix* on wetter sites. Waterlogging enhanced the uptake of Mn by leaf tissue of *E. cinerea;* both species accumulated significantly more Fe in shoot and root tissue from waterlogged soils than from the control treatment.

Kothny (1973) reported that the plant availability of Hg is decreased in reducing soils. Suggested mechanisms of Hg retention by soils under anaerobic conditions included insoluble complex formation with soil organics and sulfide.

Mikkelsen and Brandon (1975) reported that Zn deficiency in rice is observed to be more severe in submerged soils than in surface irrigated soils, especially when crop residues are undergoing active decomposition. Though no direct reference is made to the effects of flooding and residue decomposition on soil redox potential, it is probable that the Zn deficiency problems noted were related to the oxidation-reduction conditions of these soils.

Jugsujinda (1975) found that rice uptake of Fe, Mn, and Zn was strongly influenced by redox potential and pH in a laboratory study in which these parameters were controlled. Labeled Fe uptake was high under anaerobic conditions at low pH. Labeled Mn uptake was considerably greater under reduced conditions at both pH 5.5 and 7.5. Zinc uptake was reduced as pH increased in incremental units from 5.0 to 8.0. In contrast to Fe and Mn, Zn uptake was greater in aerobic soil than in anaerobic soil at each pH level studied.

In a literature review, Lucas and Knezek (1972) reported that mixed effects of poor aeration on Zn availability to plants have been found. However, much of the available literature indicates Zn is less available to plants in anaerobic soils (Mikkelsen and Brandon 1975; Jugsujinda 1975).

Kirkham (1975) reported that Cd accumulation in barley roots grown on sludge-amended soil was inversely proportional to the Cd retained in sludge crusts, which, in turn, depended on sludge application frequency. Though no redox potential measurements were made, processes affected by redox potential seemed to be involved as it was stressed that application frequency strongly influenced the wetting and drying cycles in the upper soil layer and sludge crust.

Bingham et al. (1976) reported on the gamma-leaf content of Cd, Cu, Fe, Mn, and Zn in rice grown under flooded and nonflooded conditions in a soil amended with 1% sewage sludge enriched with variable levels of Cd. Grain yield was reduced at 17 μg Cd g^{-1} under nonflooded conditions, while 320 μg Cd g^{-1} did not adversely affect grain yield under flooded conditions. Leaf Cd levels were slightly greater under nonflooded management. In this study, the leaf contents of Fe and Zn were found to be higher by a factor of about 0.5 to 2 under flooded conditions, while flooding increased Mn content many fold. The leaf Cu content was little affected by soil flooding.

These authors attributed the increased plant tolerance and reduced availability of Cd in flooded soils to precipitation of cadmium sulfide. However, they do not explain why sulfide precipitation failed to reduce Zn availability under flooded conditions. Gambrell et al. (1977a), working with four river and coastal sediments, reported that the chemical availability of Cd is enhanced by sequential oxidation of sediment-water systems above the critical potential for sulfide stability. Increased Cd release from sediment organics with increasing soil oxidation intensity apparently contributed to the higher levels of Cd in soluble and exchangeable forms in the Gambrell et al. study.

Ito and Iimura (1975) reported that the Cd content of rice plants increased only slightly with increasing Cd additions to submerged soils in pot experiments. However, under oxidizing conditions, the Cd content of rice plants increased markedly with Cd additions. Cadmium concentration in the soil solutions ranged from approximately 0.1 μg Cd l^{-1} in the submerged soil to 10 μg Cd l^{-1} in the drained soil.

Reddy and Patrick (1977) reported that an increase in the redox potential of a Mhoon soil suspension increased the uptake of Cd by lowland rice to a considerable extent.

In a study of the effects of controlled pH and oxidation-reduction conditions on trace and toxic metal uptake by marsh plants, Gambrell et al. (1977b) reported that Cd and Zn uptake were enhanced by an increase in soil oxidation levels.

From the literature, it is apparent that the various processes affecting metal availability in soils are influenced by both redox potential and pH. A study in which both of these parameters are controlled at specified levels covering the ranges normally encountered in soils would be helpful in understanding how soil conditions affect the plant availability of metals applied with sludge. Currently, little information is available in the literature on the effect of pH on this topic, and essentially no information has been published regarding the influence of soil redox potential on mobilization of sludge-bound toxic heavy metals.

1.3 Objectives and Experimental Approach

The primary objective of this study was to determine the effects of soil pH and especially oxidation-reduction (redox) conditions on Cu, Zn, and Cd levels in various soil chemical forms, and on the relative plant availability of these trace and toxic metals in agricultural soils receiving municipal sludge applications.

In the chemical availability studies, sludge-amended, soil-water suspensions were incubated under conditions of controlled pH and redox potential over the range of these parameters expected in field situations. The effects of these parameters on transformations between readily available and potentially available chemical forms were determined by chemical extractions performed on suspension aliquots from each treatment.

The effects of pH and redox potential on plant availability of these metals were determined by growing corn and rice seedlings for 2 weeks in sludge-amended soil suspensions in which pH and redox potential were controlled.

2 Materials and Methods

2.1 Sludge and Soil Sources

2.1.1 Chicago Reference Sludge

Chicago reference sludge material was obtained from J. Davidson of the University of Florida at Gainesville. Grinding was the only additional preparation required before using this material in the laboratory microcosms.

2.1.2 Drummer Silty Clay Loam Soil Material

The Drummer silty clay loam soil material was obtained from R. Pope, Agricultural Extension Service, University of Illinois. This material was finely grounded for use in the various laboratory studies.

2.1.3 Loring Silt Loam Soil Material

A Loring silt loam soil material was collected from a wooded area within Baton Rouge. Classification was based on soil survey maps (Dance et al. 1968). The surface (A) horizon collected was dried, ground, and well mixed for subsequent laboratory studies.

2.2 Preliminary Study – Effect of Air or N_2 Purging on Extractable Cu, Zn, and Cd in Soil/Sludge Mixtures of Various Ratios

A preliminary study was conducted in which mixtures of Chicago reference sludge and Drummer soil were incubated as suspensions under conditions of continuous air purging or oxygen-free N_2 purging. The sludge/soil ratios used were 0, 5, and 25%. At the end of a 2-week incubation, the suspensions were sequentially extracted for metals in the exchangeable and DTPA extractable forms using the procedures described elsewhere in this work.

2.3 Incubation of Sludge: Soil Suspensions Under Controlled pH and Redox Potential Conditions

A small (approximately 4-liter) desiccator base was used to contain the aqueous suspensions of sludge-soil mixtures incubated under controlled pH and redox potential conditions (Fig. 1). Thirty g of sludge, 270 g of soil, and 1500 ml of deionized water were placed in each desiccator to achieve a water-to-solids ratio of 5:1.

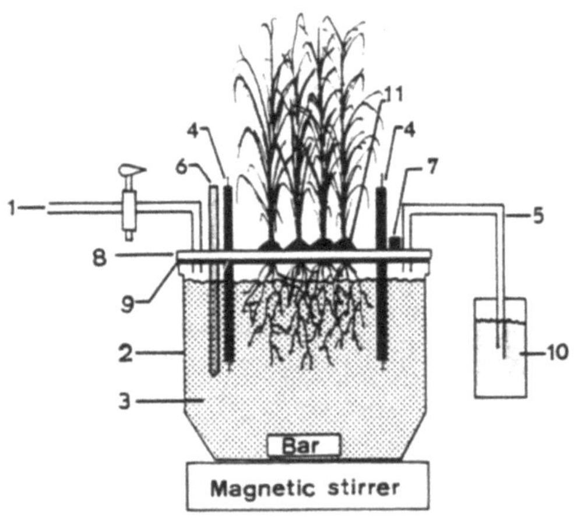

1. Gas inlet
2. Desiccator base
3. Soil suspension
4. Platinum electrode
5. Gas outlet
6. Salt bridge
7. Serum cap
8. Plexiglass plate
9. Closed-cell foam rubber seal
10. Distilled water
11. Permagum seal

Fig. 1. Experimental microcosm. Bright platinum electrodes were collected to a Beckman Zeromatic IV pH/millivolt meter for measuring redox potential; a saturated calomel reference electrode connected to the suspension with a salt bridge completed the electrochemical cell. Automatic control of redox potential was achieved by connecting a meter relay (General Electric, Type 196) to the recorder output of the millivolt meter. The meter relay activated an air pump whenever the redox potential of the suspension decreased below a present level. The air-flow rate was regulated between 0.5 and 2.0 ml min^{-1} to give slow oxidation of the suspension during the add-air cycle. Once the suspension returned to the preset redox potential level, the relay switched off the air pump. Oxygen-free N_2 gas was continuously purged through the suspensions to flush excess oxygen from the system at the end of the aeration cycle, and to prevent a buildup of gaseous decomposition products. Using this system, the redox potential could be maintained within 10 to 20 mV of the desired potential

A plexiglass plate was sealed over the top of the desiccator. Stirring was accomplished by means of a polyethylene-coated, four-blade stirrer which was passed through a 1-cm glass tube fixed in the desiccator cover. To prevent the diffusion of air into the suspension between the stirring rod and the glass tubing, a hole slightly smaller than the plastic shaft of the stirrer was bored in a rubber septum. This septum was placed over the end of the glass tubing with the shaft protruding from it. Glycerine was added to the top of the serum caps as a lubricant to maintain the air-tight seal between the septum and stirring rod shaft. The stirring shaft was driven by an electric, direct-drive, variable speed mixer. Each incubation vessel was equipped with a glass combination electrode for measuring pH, two platinum electrodes for measuring redox potential, a thermometer, a large rubber septum, separate inlets for air and oxygen-free nitrogen, and an outlet tube, the end of which was submerged in water to prevent atmospheric oxygen from diffusing into the flask.

Suspension pH was adjusted to the desired level and then maintained by using a syringe to add 2 *N* sodium hydroxide or hydrochloric acid, as required, through a second rubber septum. A combination pH electrode equipped with an annular (encircling the electrode body) ceramic junction for the reference electrode was found to be very stable for these studies (required infrequent recalibration). In contrast, electrodes with a small ceramic dot-type junction quickly clogged with fine clay particles or metal sulfides contributing to high and changing junction potentials.

2.4 Extraction Procedures for Chemical Availability Studies

2.4.1 General

Five selective chemical extractions were used sequentially to determine the effects of experimental variables at the levels of trace and toxic metals in specific chemical forms: (1) water-soluble, (2) exchangeable, (3) reducible, (4) organic-bound recovered by a chelating agent, and (5) residual-organic and sulfide-bound. The exchangeable fraction was a 1 *N* ammonium acetate extraction. The reducible fraction represents metals primarily associated with hydrous oxides of iron. The extractant for this fraction was 0.15 *M* oxalic acid/0.25 *M* ammonium oxalate. An 0.05 *M* solution of diethylenetriaminepenta acetic acid (DTPA) was used to extract a portion of the metals associated with the large molecular weight organic fraction. The residual organic fraction was extracted by heating with concentrated nitric acid and 30% H_2O_2. These extractions were made 2 weeks after achieving the desired pH and redox potential conditions. Details of these extraction procedures are provided elsewhere (Gambrell et al. 1977a).

It has been reported that levels of soluble and exchangeable metals in anaerobic systems can be rapidly and substantially altered by even brief exposure of the water or extract solution to air. To preserve sample integrity and eliminate experimental artifacts due to sample processing, it was necessary to conduct certain extractions of reduced samples under an inert atmosphere. This procedure is also described by Gambrell et al. (1977a).

2.5 Plant Uptake Studies

Plant uptake studies were conducted following the chemical availability studies. Plant seedlings were introduced to the incubation vessels after sampling for the chemical availability studies, and grown for 2 weeks to examine pH, and especially redox potential effects, on the availability of trace and toxic metals to plants grown in the sludge-amended soil material.

Corn and rice seeds were germinated in petri dishes filled with water (the water was changed twice daily). After germination, the seedlings were transferred to plastic cups filled with moist sand and watered twice daily, once with nutrient solution and once with water. A drain near the bottom of the cup prevented water logging and mold growth. Once the shoots reached a height of approximately 5 cm, the seedlings were fixed with electrical grade silicon sealant into plexiglass plates with the roots suspended in pots of a nutrient solution. The plants were grown under incandescent and fluorescent lighting. When the seedlings reached a height of between 10 and 20 cm and the desired soil pH and redox potential conditions were attained, the plexiglass plates in which the plants were mounted were substituted for those currently in place on the experimental microcosms. The various electrodes, stirring assembly, and gas tubes were fitted into the new plate and sealed. Now the plant roots were immersed in the sludge-amended soil suspensions maintained under controlled pH and redox potential conditions. The plants were grown in the sludge-soil suspension for a 2-week period under mixed incandescent and fluorescent lighting. After 2 weeks, the stem and leaf tissues were harvested, dried at 70 °C, cut into small pieces with stainless steel scissors, weighed, and digested in hot, distilled nitric acid for metal analyses. Data are presented for two soil materials amended with the Chicago reference sludge.

All soil extract and plant tissue digests were analyzed for Cu, Zn, and Cd on an inductively coupled argon plasma emission spectrometer.

2.6 Experimental Incubations

Mixtures of Chicago reference sludge and Drummer or Loring soil material were incubated as suspensions under controlled pH (5.0, 6.5 and 8.0) and redox potential (-150, 50, 250, and 500 mV) conditions where -150 mV represents strongly reducing (anaerobic) conditions and 500 mV represents well-oxidized (aerobic) soil conditions. For purposes of comparing with other chemical forms, levels of dissolved metals were reported as micrograms of metal in the aqueous phase divided by the number of grams of oven-dry soil solids in the sampled aliquot.

2.7 Effect of Aerobic or Anaerobic Sewage Digestion Method on Plant Availability of Metals

A study was conducted in which young soybean and corn plants were grown for 3 weeks in a Drummer soil material amended with 5% sludge derived from either aerobic or anaerobic digestion of a fresh Baton Rouge sewage aliquot. Raw sewage was collected from the Baton Rouge Central Treatment Plant, divided into equal portions, partially dewatered, then one portion was digested anaerobically and the other aerobically. The above-ground plant tissue was harvested and analyzed for metals. Each treatment was replicated five times.

3 Results and Discussion

3.1 Chemical Availability Studies – Preliminary Study with Air and N_2 Purged Sludge/Soil Suspensions

A preliminary study on soil oxidation effects on extractable levels of Cu, Zn, and Cd using simple air or N_2 purging of soil/sludge suspensions was conducted. Increasing the ratio of sludge to soil increased the levels of metals extracted, and, at high ratios of sludge, substantially more Cu, Zn, and Cd were extracted in potentially mobile and plant-available forms (Table 1). This simple study suggested the need for a more detailed study of redox potential effects on chemical forms and plant availability of metals in sludge-amended soils.

3.2 Chemical Availability Studies – Controlled pH and Redox Potential

3.2.1 Cu

Table 2 presents data on the effect of pH and redox potential on levels of Cu in the various chemical forms studied. There was a trend toward increased dissolved, exchangeable, and DTPA-extractable forms with increasing redox potential. Dissolved and exchangeable Cu represent the most readily available forms, while the DTPA-extractable form most likely represents potentially available Cu that is held to soil organic matter by retention mechanisms intermediate in strength between the exchangeable form and the H_2O_2 digestible form. The bonding strength of this type of organic matter association apparently decreases with increasing redox potential, as indicated by the greater recovery of DPTA-extractable Cu under oxidized conditions.

One of the most striking changes in levels of Cu as a function of changing redox potential is the sequential increase in reducible Cu levels with increasing redox potential. This is believed to be Cu associated with the hydrous iron oxide fraction. Reducible Cu accounted for most of the potentially available Cu under

Table 1. The effect of the sludge-soil ratio and soil oxidation conditions on exchangeable and DTPA-extractable levels of Cu, Zn, and Cd in a suspension of Chicago reference sludge and Drummer soil

Metal	Sludge/soil ratio (%)	Purge conditions	Chemical form ($\mu g\ g^{-1}$ soil)	
			Exchangeable	DTPA-extractable
Cu	–	Nitrogen	nd[a]	5.5[b]
Cu	–	Air	1.08	8.5
Cu	5	Nitrogen	nd	1.5
Cu	5	Air	1.00	48.5
Cu	25	Nitrogen	nd	6.5
Cu	25	Air	3.20	167.0
Zn	–	Nitrogen	1.34	20
Zn	–	Air	1.64	23
Zn	5	Nitrogen	0.20	32
Zn	5	Air	10.76	143
Zn	25	Nitrogen	1.36	210
Zn	25	Air	21.80	400
Cd	–	Nitrogen	0.14	0.35
Cd	–	Air	0.04	0.16
Cd	5	Nitrogen	nd	0.52
Cd	5	Air	1.90	7.70
Cd	25	Nitrogen	0.12	4.52
Cd	25	Air	6.06	27.70

[a] Less than detection limits.
[b] Average of two replications.

oxidizing conditions. The residual organic bound (H_2O_2 digestible) Cu decreased sequentially with increasing oxidation conditions. Under strongly reducing conditions, such as −150 mV, the H_2O_2 digestion will also recover metals precipitated as insoluble sulfides.

It is apparent from these data that the organic (+sulfide) form of Cu predominates under reducing conditions, and that a large proportion of this potentially available Cu is transformed to a predominately mineral form (reducible) under oxidizing conditions. Although relatively large quantities of Cu are moving between the potentially available forms as redox potential changes, relatively small shifts occur in the amounts of the readily available soluble and exchangeable forms.

3.2.2 Zn

Dissolved and exchangeable Zn levels were much greater at pH 5.0 than at the two higher pH levels in the Chicago sludge/Drummer soil mixtures (Table 3). This is

Table 2. The effect of pH and redox potential on levels of Cu in selected chemical forms in a Drummer soil/Chicago reference sludge material

		Redox potential (mV)			
pH	Form	-150	50	250	500
		($\mu g\ g^{-1}$, Oven-dry solids basis)			
5.0	Total dissolved[a]	m[b]	0.35[c]	0.55	0.55
	Exchangeable	m	0.0	3.15	3.55
	Reducible	m	3.30	102.0	75.6
	DTPA-extractable	m	3.50	19.6	22.6
	H_2O_2 extractable	m	119.0	32.0	28.0
7.0	Total dissolved	0.10	0.05	0.10	0.25
	Exchangeable	2.90	0.70	1.90	2.45
	Reducible	1.05	29.4	91.0	112.0
	DTPA-extractable	3.40	12.0	19.7	19.0
	H_2O_2 extractable	m	m	m	m
8.0	Total dissolved	0.40	0.45	0.40	0.90
	Exchangeable	0.10	2.70	2.65	3.15
	Reducible	7.15	71.5	61.8	101.0
	DTPA-extractable	4.10	15.0	23.6	20.4
	H_2O_2 extractable	136.0	64.7	52.9	32.9

[a] Total dissolved expressed as $\mu g\ g^{-1}$ oven-dry solids basis where water to solids ratio was 5:1 (w/w).
[b] Missing value.
[c] Mean of duplicate subsamples.

Table 3. The effect of pH and redox potential on levels of Zn in selected chemical forms in a Drummer soil/Chicago reference sludge material

		Redox potential (mV)			
pH	Form	-150	50	250	500
		($\mu g\ g^{-1}$, oven-dry solids basis)			
5.0	Total dissolved[a]	m[b]	34.4[c]	42.4	41.6
	Exchangeable	m	121.0	116.0	111.0
	Reducible	m	220.0	232.0	200.0
	DTPA-extractable	m	61.2	45.8	48.8
	H_2O_2 extractable	m	110.0	120.0	118.0
7.0	Total dissolved	0.40	0.80	1.35	1.50
	Exchangeable	0.90	12.4	18.5	21.0
	Reducible	308.0	316.0	266.0	283.0
	DTPA-extractable	93.6	75.7	58.1	54.0
	H_2O_2 extractable	m	m	m	m
8.0	Total dissolved	0.75	0.60	0.75	0.50
	Exchangeable	14.9	21.8	22.6	20.0
	Reducible	303.0	329.0	224.0	269.0
	DTPA-extractable	67.2	77.8	84.6	60.0
	H_2O_2 extractable	132.0	130.0	182.0	183.0

[a] Total dissolved expressed as $\mu g\ g^{-1}$ oven-dry solids basis where water to solids ratio was 5:1 (w/w).
[b] Missing value.
[c] Mean of duplicate subsamples.

in accord with the commonly observed effects of pH on soluble soil Zn levels and plant availability as reported in the literature. There was a weak trend for these readily available Zn forms to increase with increasing redox potential at the lower pH levels.

Most of the total potentially available Zn present was in the reducible form with the residual organic bound (H_2O_2 digestible) and DTPA-extractable levels removing approximately one-half and one-fourth the reducible levels, respectively.

3.2.3 *Cd*

Compared to Cu and Zn, a much larger proportion of the potentially available Cd present was recovered in the dissolved and exchangeable forms (Table 4). This relatively greater chemical availability is reflected in the frequently reported excess concentrations of Cd found in plant tissue grown on Cd-contaminated soils.

In comparison to most other trace and toxic metals, Cd appears to be less strongly associated with organics, and is particularly little retained by hydrous oxides of Fe (reducible phase). In other studies designed to specifically examine this association, Kinniburgh et al. (1977) and Gambrell et al. (1977a) also found Cd to be weakly bound to hydrous oxides of Fe compared to other metals. Thus,

Table 4. The effect of pH and redox potential on levels of Cd in selected chemical forms in a Drummer soil/Chicago reference sludge material

pH	Form	Redox potential (mV)			
		-150	50	250	500
		($\mu g\ g^{-1}$, oven-dry solids basis)			
5.0	Total dissolved[a]	m[b]	1.59[c]	2.28	2.08
	Exchangeable	m	7.27	7.50	7.48
	Reducible	m	2.40	3.53	2.69
	DTPA-extractable	m	2.37	3.16	3.09
	H_2O_2 extractable	m	8.30	7.10	6.70
7.0	Total dissolved	0.01	0.08	0.12	0.13
	Exchangeable	0.0	2.96	3.52	3.96
	Reducible	0.42	3.14	3.26	3.11
	DTPA-extractable	1.66	2.16	2.88	2.23
	H_2O_2 extractable	m	m	m	m
8.0	Total dissolved	0.08	0.08	0.06	0.06
	Exchangeable	0.02	2.97	2.94	2.66
	Reducible	2.95	3.70	2.52	3.22
	DTPA-extractable	3.26	3.31	4.20	3.17
	H_2O_2 extractable	15.8	12.5	12.4	13.2

[a] Total dissolved expressed as $\mu g\ g^{-1}$ oven-dry solids basis where water to solids ratio was 5:1 (w/w).
[b] Missing value.
[c] Mean of duplicate subsamples.

this research supports other studies which indicate that Cd contamination of soils and sediments represents a greater environmental risk than many other toxic metals at equivalent levels because of its relatively high mobility. Results for the chemical availability studies of Cd are given in Table 4.

As will be discussed elsewhere in this work, the Cd levels in plant tissue, like Zn, increased greatly with increasing redox potential levels at all pH levels. Levels of dissolved Cd responded similarly only at pH 7 in this chemical availability study, while exchangeable Cd increased markedly between -150 and 50 mV at both of the two higher pH levels studied.

3.3 Plant Uptake Studies

After sludge-soil suspension aliquots were withdrawn for the chemical availability studies described above, either rice or corn seedlings were transplanted into the suspensions and grown for a 2-week period. Plant material from each experimental unit was harvested as two subsamples which were digested and analyzed separately. Tables 5 and 6 give the mean Cu, Zn, and Cd concentrations for subsamples of rice and corn seedlings grown for 2-week periods under controlled pH and redox potential conditions.

These plant uptake studies were intended to indicate relative treatment effects of varying soil physicochemical conditions on plant availability of the selected trace and toxic elements. It would be difficult to compare the plant concentration data obtained with other plant concentration data in the literature because of the short (2 week) growing period and because the plants were harvested at a very early age. Therefore, it is believed that the data in Tables 5 and

Table 5. Effect of pH and redox potential on concentrations of selected elements in rice grown in sludge-soil suspensions

Sludge/ soil	Element	pH	Redox Potential (mV) -150	50	250 (Metal, $\mu g\ g^{-1}$)	500
C/D[a]	Cu	5.0	15	30	32	30
C/D	Cu	6.5	102	105	71	84
C/L[b]	Cu	6.5	58	46	71	30
C/L	Cu	8.0	20	36	22	24
C/D	Zn	5.0	81	110	115	264
C/D	Zn	6.5	158	150	238	233
C/L	Zn	6.5	71	65	147	173
C/L	Zn	8.0	29	26	62	85
C/D	Cd	6.5	2.4	2.4	12	11
C/L	Cd	6.5	1.8	2.9	28	53
C/L	Cd	8.0	0.5	1.5	2.0	2.6

[a] Chicago reference sludge/Drummer soil material.
[b] Chicago reference sludge/Loring soil material.

Table 6. Effect of pH and redox potential on concentrations of selected elements in corn grown in sludge-soil suspensions

Sludge/ soil	Element	pH	Redox potential (mV) -150	50	250 (Metal, $\mu g\ g^{-1}$)	500
C/D[a]	Cu	5.0	m[b]	19	14	37
C/D	Cu	6.5	25	22	29	42
C/D	Cu	8.0	13	17	19	32
C/D	Zn	5.0	m	143	98	203
C/D	Zn	6.5	103	184	298	925
C/D	Zn	8.0	85	187	245	563
C/D	Cd	5.0	m	5.7	7.7	12
C/D	Cd	6.5	2.1	6.2	24	110
C/D	Cd	8.0	0.3	0.8	8.0	12

[a] Chicago reference sludge/Drummer soil.
[b] Missing data for one or both subsamples.

6 can best be used by focusing on relative redox potential effects within a given sludge/soil mixture and pH combination. Also, for the pH 5.0 corn study, there was some concern that the plants may have been stressed, either by a toxicity effect or another factor, as these metals were expected to be present in the highest concentration in the pH 5.0 treatment based on literature reports of pH effects on metal uptake and similar work conducted in this lab with marsh plants and dredged materials. However, metal levels in the pH 5.0 treatment were almost always lower than metal concentrations in the pH 6.5 treatment.

Results with rice were mixed, but increasing oxidation conditions tended to increase Cu in corn.

Of the elements summarized in Tables 5 and 6, both Zn and Cd responded markedly to redox potential conditions. For corn at pH 6.5 and 8.0, Zn levels increased from around 100 ppm under strongly reducing conditions to greater than 900 and 500 ppm at pH 6.5 and 8.0, respectively, in well-oxidized, sludge-soil mixtures.

Cadmium concentrations also increased greatly as the redox potential increased from strongly reduced to well-oxidized conditions in most combinations of sludge-soil mixtures and pH. For example, a 20-fold or greater increase occurred between -150 mV and 500 mV in rice grown at pH 6.5 in the Chicago sludge/Loring soil suspension and in the pH 8.0 Chicago sludge/Drummer soil combination. A 50-fold increase occurred between the redox potential extremes in the Chicago sludge/Drummer soil combination incubated at pH 6.5. Gambrell et al. (1980) and Folsom et al. (1981) reported similar redox potential effects on Cd uptake by marsh plants. There were many relatively large sequential increases in plant tissue Cd (and Zn) concentrations between adjacent redox potential levels, which suggests that even modest changes in soil oxidation-reduction conditions can greatly affect plant Zn and Cd concentrations. This can be a very important consideration in managing Cd and Zn contaminated soils.

3.4 Effect of Aerobic or Anaerobic Sewage Digestion Method on Plant Availability of Metals

The effects of aerobic and anaerobic digestion of Baton Rouge sewage on the concentration of Cu, Zn, and Cd in soybean and corn plants grown on a 5% sludge-amended Drummer soil material is shown in Table 7. In soybeans, neither the presence of sludge nor the sewage digestion method had any significant effect on Cu, Zn, and Cd in the plant tissue. For corn, the presence of the sludge significantly increased Cd content of the above-ground tissue, but there was no significant digestion treatment effect on Cd levels. The anaerobically digested sludge resulted in a significantly higher level of Cu in the corn, but the increase was relatively small and well within levels of Cu normally found in plant tissue.

4 Summary and Conclusions

A laboratory study was conducted using sludge-amended soils to examine the effects of pH and especially redox potential (oxidation-reduction conditions) on the levels of Cu, Zn, and Cd in various inorganic and organic soil chemical forms. Also, the effect of redox potential on the availability of these elements to plants was determined. A preliminary study had indicated either air purging or N_2 purging various soil/sludge ratios had a marked effect on the extractable levels of these metals.

Various organic phases predominate in retaining Cu under reducing conditions, but as redox potential increased, a very marked transformation occurred to the reducible (hydrous iron oxide) phase. Potentially available Cu was most strongly associated with the iron oxide phase under well-oxidized conditions. There was a trend for dissolved, exchangeable, and chelate-extractable Cu to

Table 7. Cu, Zn, and Cd content in soybean and corn plants grown in a 5% aerobically and anaerobically digested Baton Rouge sludge/drummer soil mixture

Plant	Digestion treatment	Element		
		Cu	Zn ($\mu g\ g^{-1}$)	Cd
Soybean	Control (0%)	12.9a[a]	67.5a	0.17a
Soybean	Anaerobic (5%)	13.7a	75.6a	0.14a
Soybean	Aerobic (5%)	12.7a	75.4a	0.16a
Corn	Control (0%)	14.7a	147a	0.10a
Corn	Anaerobic (5%)	20.8b	154a	0.45b
Corn	Aerobic (5%)	15.6a	135a	0.47b

[a] Means of five replications. Means followed by the same letter within a plant/metal combination are not significantly different at the 5% level using Duncan's Multiple Range Test.

increase with increasing redox potential. Results with rice were mixed, but increasing oxidation conditions tended to increase Cu in corn.

In the Chicago sludge-Drummer soil studies, there was a weak trend for readily available (dissolved and exchangeable) Zn to increase with increasing redox potential, and the highest levels were at pH 5.0, as expected. A very large increase in plant Zn levels occurred with increasing redox potential. Most of the potentially available Zn was found in the reducible fraction under oxidized and reduced conditions.

Compared to Cu and Zn, a much larger proportion of the total Cd extracted was recovered in the most readily available chemical forms (dissolved and exchangeable). Cadmium appeared less strongly associated (immobilized) with high molecular weight organics, and, in particular, was weakly associated with hydrous iron oxides. This contributes to a situation where Cd is readily accumulated by plants grown in Cd-contaminated soils. In most of the experimental combinations of sludge-soil mixtures and pH used in the laboratory microcosms, plant Cd concentrations increased greatly as the redox potential increased.

A fresh Baton Rouge sewage effluent was divided into two aliquots and subjected to aerobic and anaerobic digestion. These treatments had little effect on metals in young soybean plants grown in a 5% sludge/Drummer soil mixture. Copper in corn may have been slightly elevated in the anaerobic digested sludge/soil mixture compared to the aerobic digestion treatment, but the sludge digestion method did not affect Zn and Cd levels in corn.

Of particular interest, this research demonstrated the important effects of soil redox potential conditions in regulating the chemical mobility and plant availability of Zn and Cd, two key contaminants in sludge materials that may be applied to soils. Obviously, most crop species are not grown on soils that range from strongly reduced to well oxidized. Most crops are grown on well-drained (oxidized) soils. Rice is an important exception in that it is normally grown on flooded (reduced) soils. But, rice may be grown in upland, oxidized soils as well, and in some regions of the country, substantial acreage is devoted to "dry-land" rice. Sugarcane is grown on soils that are somewhat low and wet, and the redox potentials on these soils would probably be intermediate, but subject to frequent and wide fluctuations. Nonagricultural soils may also be used to dispose of sewage sludge, and the oxidation status encountered in various natural habitat soils or other uncultivated soils may vary considerably.

In considering the potential problems associated with the disposal or productive use of contaminated sewage sludges, there are many factors and options to be considered in developing acceptable management plans. This research indicates that the oxidation-reduction status of soils being considered for receiving municipal sludge is an important consideration, particularly for Zn and Cd. In those situations where feasible alternatives exist, wet or poorly oxidized soils would be more effective in immobilizing Zn and Cd since application to these soils would result in less plant accumulation of these metals for most crops, or plants in uncultivated habitats.

Acknowledgments and disclaimer. Although the research presented in this manuscript has been funded wholly or in part by the U.S. Environmental Protection Agency under Contract No. R805789 to Louisiana State University, it has not been subjected to agency review and therefore does not necessarily reflect the views of the U.S. EPA and no official endorsement should be inferred. The mention of any trade names does not imply product endorsement by the U.S. EPA or Louisiana State University.

References

Bingham FT, Page AL, Mahler RJ, Ganje TJ (1976) Cadmium availability to rice in sludge-amended soil under 'Flood' and 'Nonflood' cultures. Soil Sci Soc Am J 40:715–719

Braids OC, Ardakani MS, Molina JAE (1970) Liquid digested sewage sludge gives field crop necessary nutrients. Ill Res 12 (3):6–7

Clark F, Nearpass DC, Sprecht AW (1957) Influence of organic additions and flooding on iron and manganese uptake by rice. Agron J 48:586–589

Dance RE, Griffis BJ, Nutt BB, White AG, Lytle SA, Seaholm JE (1968) Soil survey of East Baton Rouge Parish, Louisiana. Soil Cons Serv, US Dep Agric, and Louisiana Agric Exp Stn

Engler RM, Patrick WH Jr (1975) Stability of sulfides of manganese, iron, zinc, copper, and mercury in flooded and nonflooded soil. Soil Sci 119:217–221

Folsom BL Jr, Lee CR, Bates DJ (1981) Influence of disposal environment on availability and plant uptake of heavy metals in dredged material. Tech Rep EL-81-12, Environ Lab, US Army Eng Waterways Exp Stn, Vicksburg, Miss, 151 pp

Gambrell RP, Khalid RA, Verloo MG, Patrick WH Jr (1977a) Transformations of heavy metals and plant nutrients in dredged sediments as affected by oxidation-reduction potential and pH. Part II. Materials and methods, results and discussion. Contract DACW-74-C-006, Off Dredged Mater Res, US Army Eng Waterways Exp Stn, Vicksburg, Miss, 315 pp

Gambrell RP, Collard VR, Reddy CN, Patrick WH Jr (1977b) Trace and toxic metal uptake by marsh plants as affected by oxidation-reduction conditions, pH, and salinity. Contract DACW-39-73-C-0108, Off Dredged Mater Res, US Army Eng Waterways Exp Stn, Vicksburg, Miss, 135 pp

Gambrell RP, Khalid RA, Patrick WH Jr (1980) Chemical availability of mercury, lead, and zinc in Mobile Bay sediment suspensions as affected by pH and oxidation-reduction conditions. J Environ Sci Technol 14:431–436

Ito H, Iimura K (1975) Absorption of cadmium by rice plants in response to change of oxidation-reduction conditions of soils. Abst Nippon Rozyohiryogaku Zasshi (J Sci Soil Man) Japan 46:82–88

Jenne EA (1968) Controls of Mn, Fe, Co, Ni, Cu, and As concentrations in soils and water: the significant role of hydrous Mn and Fe oxides. Trace inorganics in water. Adv Chem Ser 73:337–388

Jones HE, Etherington JR (1970) Comparative studies of plant growth and distribution in relation to waterlogging. 1. The survival of *Erica cinera* L. and *E. tetralix* L. and its apparent relationship to iron and manganese uptake in waterlogged soil. J Ecol 58:487–496

Jugsujinda A (1975) Growth and nutrient uptake by rice under controlled oxidation-reduction and pH conditions in a flooded soil. PhD Diss, Dep Agron, La State Univ, Baton Rouge

Kinniburgh DG, Sridhar K, Jackson ML (1977) Specific adsorption of zinc and cadmium by iron and aluminum hydrous oxides. In: Drucker H, Wildung RE (eds) Biological implications of metals in the environment. ERDA Symp Ser 42, Energ Res Dev Admin, Oak Ridge, Tenn pp 231–239

Kirkham MB (1975) Trace elements in corn grown on long-term sludge disposal site. Environ Sci Technol 9:765–768

Kothny EL (1973) The three-phase equilibrium of mercury in nature. In: Kothny EL (ed) Trace elements in the environment. Adv Chem Ser Am Chem Soc, Washington, DC 123:48–80

Krauskopf KB (1972) Geochemistry of micronutrients. In: Micronutrients in agriculture. Soil Sci Soc Am Inc, Madison, Wisc, pp 7–40

Lee GF (1973) Role of hydrous metal oxides in the transport of heavy metals in the environment. Proc Symp Heavy metals in the environment, Vanderbilt Univ, pp 137–147

LeLand HV, Shukla S, Shimp NF (1973) Factors affecting distribution of lead and other trace elements in sediments of southern Lake Michigan. In: Singer PC (ed) Trace metals and metal-organic interactions in natural waters. Ann Arbor Science Publishers, Inc, Ann Arbor, Michigan pp 89–129

Loganathan AP, Burau RG, Fuerstenau DW (1977) Influence of pH on the sorption of Co^{2+}, Zn^{2+} and Ca^{2+} by a hydrous manganese oxide. Soil Sci Soc Am J 41:57–62

Lucas RE, Knezek BD (1972) Climatic and soil conditions promoting micronutrient deficiencies in plants. In: Dinauer RC (ed) Micronutrients in agriculture. Soil Sci Soc Am Inc, Madison, Wisc pp 265–288

Mandal LN (1962) Levels of iron and manganese in soil solution and the growth of rice in waterlogged soils in relation to the oxygen status of soil solution. Soil Sci 94:387–391

Mikkelsen DS, Brandon DM (1975) Zinc deficiency in California rice. Cal Agric vol 29:9:8–9

Morel F, McDuff RE, Morgan JJ (1973) Introductions and chemostasis in aquatic chemical systems: role of pH, pE, solubility, and complexion. In: Trace metals and metal-organic interactions in natural waters. Ann Arbor Sci, Mich

Peterson JR, McCalla TM, Smith GE (1971) Human and animal wastes as fertilizers. In: Fertilizer technology and use. Soil Sci Soc Am Inc, Madison, Wisc, pp 557–596

Reddy CN, Patrick WH Jr (1977) Effect of redox potential and pH on the uptake of Cd and Pb by rice plants. J Environ Qual 6:259–262

Schnitzer M, Skinner SIM (1966) Organo-metallic interactions in soils: 5. Stability constants of Cu^{++}-, Fe^{++}-, and Zn^{++}-fulvic acid complexes. Soil Sci 102:361–365

Stevenson FJ, Ardakani MS (1972) Organic matter reactions involving micronutrients in soils. In: Micronutrients in agriculture. Soil Sci Soc Am Inc, Madison, Wisc, pp 79–114

Taylor RM, McKenzie RM (1977) The association of trace elements with manganese minerals in Australian soils. Aust J Soil Res 4:29–35

Verloo M, Cottenie A (1972) Stability and behavior of complexes of Cu, Fe, Mn, and Pb with humic substances of soils. Pedologie 22:174–184

Weeraratna CS (1969) Adsorption of manganese by rice under flooded and unflooded conditions. Plant Soil 30:121–125

Windom HL (1973) Investigation of changes in heavy metals concentrations resulting from maintenance dredging of Mobile Bay Ship Channel, Mobile Bay, Alabama. Rep submitted to US Army Corps of Engineers, Mobile District, Al

Microbiological Aspects of Heavy Metal and Toxic Chemical Behaviour in Porous Media

K.H. Domsch[1]

1 Introduction

Potentially toxic inorganic and organic chemicals reach the soil by either direct application or deposition of emissions or as contaminants of plant litter. Although 65 elements with a specific gravity > 5 can be considered as heavy metals, only about 25% of them have been included in ecotoxicity studies. This is because the remaining elements are either extremely rare or insoluble and the environmental impact is considered to be small (Duxbury 1985). The number of organic chemicals which require environmental monitoring, however, is considerably higher. Pesticides designed to interfere with living cells and used in open systems are obvious examples of organic chemicals which need to be considered with priority in this context.

The information available on the behaviour of toxic chemicals in soil is highly variable, reflecting the wide variety of chemical properties, microbial responsiveness and soil conditions. Since there is no uniform chemical behaviour and no unique microbial or soil characteristic, the main emphasis in the following paragraphs will be on examples which allow the recognition of some of the more frequent patterns and also the description of ranges of microbial responses.

Microorganisms interact with chemicals in two ways: they may suffer from the chemical as a toxicant or modify the chemical as a substrate. Both aspects will be considered and in addition the environmental influences on chemical and/or microbial behaviour will be briefly discussed.

2 Microbiological Aspects

Three general observations may be made: (1) the soil microbial community is composed of organisms ranging from high tolerance to high sensitivity. When a toxic element is introduced into this community, long-term selection pressure will increase the proportion of tolerant organisms. (2) Doses (contents, concentrations) of toxic agents in the soil are rarely suitable for recognizing the true

[1]Department of Soil Biology, Federal Agricultural Research Centre, Bundesallee 50, 3300 Braunschweig, FRG

dose-response relations because the effective dose is obscured by interfering soil conditions (see below). (3) Synergistic effects must be assumed to occur but are not sufficiently studied. The information given below applies to aerobic soil conditions unless otherwise stated.

2.1 Influence on Microbial Communities

Terrestrial ecosystems in an advanced state of maturity support a quantity of microbial biomass which is closely linked to the annual flux of carbon through the system. Under mature (climax) conditions the system is characterized by a maximum of metabolic and taxonomic diversity. Toxic chemicals interfere with both. Field data illustrating the relationship between Cu concentration and quantity of microbial biomass are given in Fig. 1.

The example in Fig. 1 represents a chronic multispecies dose response to a mixture of heavy metals (with EDTA-extractable Cu as an index). The heavy metals were contained in farmyard manure (with low heavy metal contents) and sewage sludge (higher contents) applied in recent years. The variability in heavy metal contents is as wide as would be expected in field experiments, and the impact on microbial biomass is remarkably high. Since the last manure and sludge application took place 17–23 years earlier, there seems to be no evidence of recovery. This would suggest that plants responded first by a continuously reduced input of residues at elevated Cu contents which then was indirectly reflected in lower microbial biomass levels. Results from short-term assays (30 days) using three different soils showed that Cu at 100 $\mu g\ g^{-1}$ had only a negligible, direct effect on microbial biomass (Beck 1981).

The qualitative aspect of heavy metal pollution is exemplified by a field investigation on fungal diversity along a steep concentration gradient (Fig. 2).

Although the grouping of the total Cu concentration in rather wide ranges indicates that the level of resolution is not very high, it does allow for the recognition of a frequently obscured phenomenon. While the frequency of more

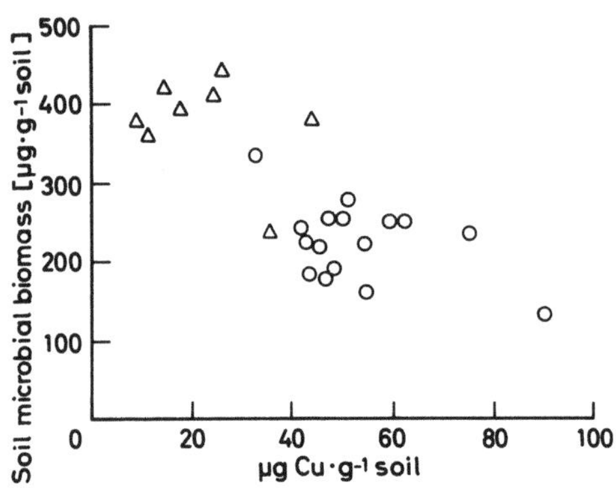

Fig. 1. Relationship between EDTA-extractable Cu and total amount of microbial biomass in soils amended with farmyard manure (△) or sewage sludge (△) (Data from Brookes et al. 1984)

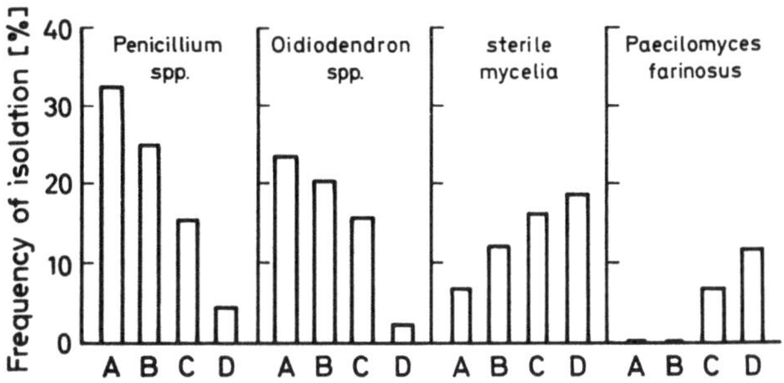

Fig. 2. Relationship between the frequency of isolation of selected fungi and total soil Cu content: *A* < 100; *B* 100–1000; *C* 1000–10 000; *D* > 10 000 μg g^{-1} (Data from Nordgren et al. 1983)

sensitive genera decreases they are replaced by other tolerant/resistant fungi. In another investigation along the same gradient the number of basidiomycetes producing fruiting bodies decreased from >40 to < 10 species (Rühling et al. 1983). Considering the role of basidiomycetes in forest floors in cycling and uptake (via mycorrhiza) of nutrients, the potential, long-term destabilizing effect of heavy metal contaminations is evident.

Another community-linked "behaviour" is the course of regeneration following the impact of a toxic chemical. Tolerance towards a chemical stress eventually develops in two ways: evolution of adapted tolerance in a slow selection process at extreme sites, and secondly a fast requisition of extra-chromosomal DNA with the corresponding R-factors. There is evidence for several elements (Hg, Cd, Sb, As) that microbial resistance is plasmid-coded but other mechanisms are also operating. It has recently been reported that as a consequence of continuing stress (Cd) and adaptation the diversity within the "new" tolerant community is partially restored (Barkay et al. 1985).

2.2 Influence on Microbial Functions

The majority of soil microorganisms depend on the annual input of reduced carbon provided by the photosynthetic products of green plants. The carbon skeleton of organic molecules is firstly utilized to satisfy the energy requirements of resting or proliferating cells and secondly to serve as substrates for the build up of microbial biomass. In this context potentially toxic chemicals may interfere with energy metabolism, catabolic or anabolic functions or the accumulation of bioelements in microbial cells.

Under normal conditions, carbon is available as "old" native soil organic matter, as recently deposited plant litter and as an organic amendment. Agricultural management tries to preserve soil organic matter but normally some will be lost due to tillage operations. The long-term application of Cu has been shown to decrease carbon loss (Mathur et al. 1979, 1980) in field experiments aimed at mitigating the loss of soil organic matter in arable soil (histosols). In this case the intended inhibitory action of the Cu would be considered a positive effect. The consequences for ecosystems which strictly depend on the functioning of nutrient cycling (e.g. forests) are negative, however. If the degradation of litter is inhibited, litter accumulates in the organic horizon and the flux of bioelements into the inorganic pool is delayed. Tyler (1976) and Coughtrey et al. (1979) have described field situations of this type in forests along Cu and Zn- or Cd-gradients respectively. The combined effect of Cu and Zn could be detected at contents of $\geq 200\ \mu g\ g^{-1}$ and that of Cd (Fig. 3) at $> 1\ \mu g\ g^{-1}$ litter.

If toxic chemicals are applied simultaneously with organic matter as is the case with sewage sludge, composted municipal waste, or even with liquid animal manure, the gross toxic effect is usually masked by the positive effect of the carbon source. These cases require long-term monitoring periods in order to recognize toxic actions. If the toxicants are not stable, however, such effects are short-lived and remain undetected.

Frequently toxic effects have been measured using soil enzymes as indicators. Unfortunately there is a lack of information on the direct effects on enzyme synthesis. In one case amylase synthesis was reported to be less sensitive than amylase activity (Cole 1977). In an attempt to condense some of the published information on heavy metal effects on enzyme activity, Fig. 4 contains a summary of results which have been obtained in short-term assays; Hg is the most toxic, while Zn, Cu, and Sn are at the end of the toxicity scale. Most of the applied doses, however, were typical for very heavily polluted soils.

The ecological significance of the inhibition of enzyme activity measured immediately following the application of a toxic chemical is rather limited since no time is allowed for recovery during short monitoring periods. In numerous experiments it has been demonstrated that even after a strong impact, enzymes resume activities close to normal within a few days. Exceptions to this are high heavy metal contents prevailing for a long time (Mathur et al. 1980).

In contrast to results from laboratory assays, the individual or combined effects of Zn and Cu on phosphatase activity under field conditions were not

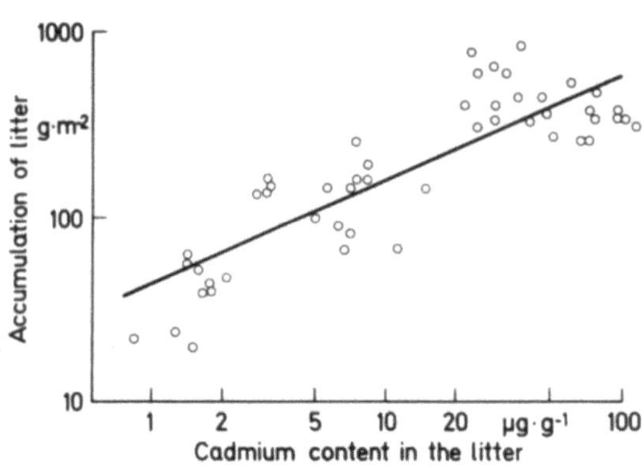

Fig. 3. Relationship between Cd content along a pollution gradient and accumulation of plant litter (Data from Coughtrey et al. 1979)

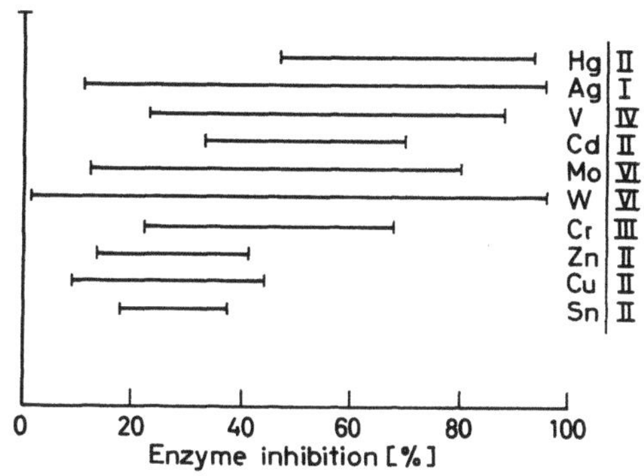

Fig. 4. Influence of heavy metal solutions on the activity of five soil enzymes: acid phosphatase (three soils at 25 μmol heavy metal g^{-1}), alkaline phosphatase (two soils at 25 μmol g^{-1}), pyrophosphatase (four soils at 10 μmol g^{-1}), arylsulfatase (four soils at 25 μmol g^{-1}), urease (six soils at 5 μmol g^{-1}) (Data from Tabatabai 1977, Juma and Tabatabai 1977, Al-Khafaji and Tabatabai 1979)

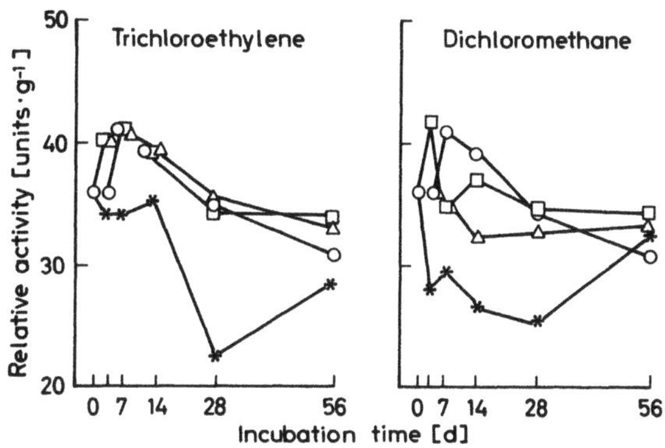

Fig. 5. Influence of volatile chlorinated hydrocarbons on phosphatase activity in closed soil-chemical-enzyme systems: ○—○ control; □—□ 0.1 μg g^{-1}; △—△ 1 μg g^{-1}; *—* 10 μg g^{-1} (Data from Kanazawa and Filip 1986)

detected below 200 μg g^{-1} (Tyler 1976, Brookes et al. 1984). Generally, the effects of heavy metals on soil enzymes are more pronounced at low pH and in soils with low sorption capacity.

Enzymes have very frequently been used as indicators for acute toxicity in pesticide side-effect assays. It is now generally accepted that even in the presence of fungicides the sensitivity of soil enzymes is rather low (Domsch in preparation).

A typical example of the behaviour of organic environmental chemicals in enzyme assays is given in Fig. 5. The frequently observed response is a temporary depression at contents which are high above expected environmental concentrations.

The accumulation of phosphorus by soil fungi and its translocation into roots is an essential function in mycorrhiza associations. In the presence of increasing heavy metal contents the following sequence of events emerges:

$<$ 5 μg [Cu, Zn, Pb]g^{-1} [soil]: no effect or stimulation of mycorrhiza;
$>$ 5 μg: no effect on mycorrhiza, increased heavy metal uptake by plants;
20–30 μg: 50% inhibition of sensitive mycorrhizas (Fig. 6);
75 μg: elimination of sensitive mycorrhizas.

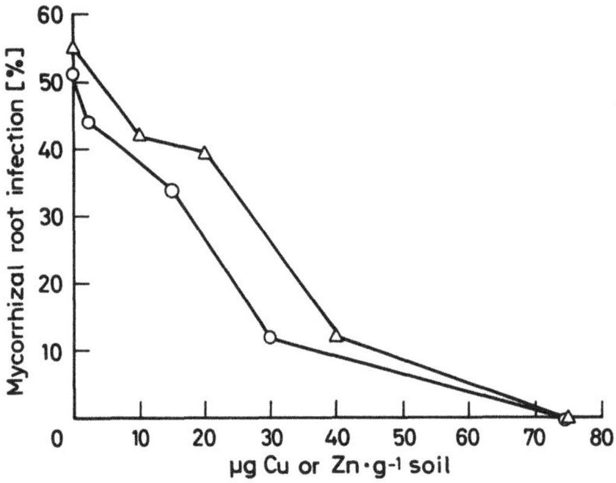

Fig. 6. Influence of Cu (○—○) or Zn (△—△) on root infection of onions by VA mycorrhiza in a sandy loam (Data from Gildon and Tinker 1983)

Table 1. Relative fungicide sensitivity of microbial indicators (Data from Domsch, in prep.)

	Sensitivity index
Ecto-mycorrhiza	4.80
Green algae	4.49
VA-mycorrhiza	3.50
Rhizobia (legumes)	2.75
Nitrifying bacteria	2.29
Degradation of organic matter	2.18

The range of toxicity may shift towards higher or lower heavy metal contents depending on soil conditions, host plant and the relative toxicity of the element. There are indications, however, that Ni is particularly toxic to mycorrhizal associations.

From the numerous reported cases it can be concluded that ectotrophic and VA-mycorrhizas belong to the group of the most sensitive indicators for non-desired side-effects of fungicides (Table 1).

2.3 Influence on Microbial Interactions

In a recent study on the ecological aspects of the heavy metal responses of microorganisms, Duxbury (1985) found it unfortunate that interactions of microbes with microbes or soil fauna "remain either completely unknown or, at best, obscure". This statement can be expanded to organic toxicants as well.

It can be assumed, however, that biological chain reactions (e.g. biomagnification in food chains, elimination of predators or antagonists) will not have the same impact on terrestrial ecosystems as they might have in aquatic systems. Physical barriers in soil usually retard or prevent widespread reactions.

The magnitude of the controlling action of protozoa, microarthropods or nematodes on soil bacteria, and actinomycetes or fungi can occasionally be observed when bacteriovore or fungivore animals are decreased or eliminated (e.g. by an insecticide or nematicide). After they stop feeding on microbes, the microbes proliferate beyond the normal level, and the apparent substrate degradation rates increase. Some of the frequently observed "stimulations" after applications of a toxic agent can be traced back to side-effects on interactions.

3 Chemical Behaviour

When a toxic agent is introduced to the biotic and abiotic components of soils, both components will determine availability, toxicity and stability of the toxic agent. Although most of the abiotic "behaviour" will be treated in other chapters of this volume, there are some microbial aspects which need to be briefly discussed in this context.

3.1 Microbial Contributions to Availability

Although heavy metals are not directly available to microorganisms as they commonly occur in nature, microbes do have a direct influence on various components of solution equilibria. Acid formation during incomplete utilization of carbohydrates, for example, particularly under anaerobic conditions, is a common phenomenon which results in increased solubilization (Fig. 7). The liberation of free Zn, Cd, Pb or Cu ions from oxides or sulphides has frequently been described for bacteria.

The mineralization of organic ligands in organo-metallic complexes (e.g. those of Cu, Pb, Ni) can also considerably increase heavy metal availability. Figure 8 gives an example of the relationship between the amount of water-soluble organic substances which are produced in the course of organic matter degradation and the Pb contents in the soil water phase.

In soils with a weak sorption capacity, microorganisms themselves take part in the immobilization of heavy metals by various accumulation mechanisms (Zamani et al. 1984). Filamentous fungi and yeast, for example, can accumulate up to 2–15 mg $[Cd]g^{-1}$ dry wt. (Trevors et al. 1986).

Finally, the microbial contributions in the transfer of heavy metals between environmental compartments must be mentioned. Some metals can be transferred from the solid to the gaseous phase by metabolic methylation or by changing the valence state (e.g. $Hg^{2+} \rightarrow Hg^{\circ}$).

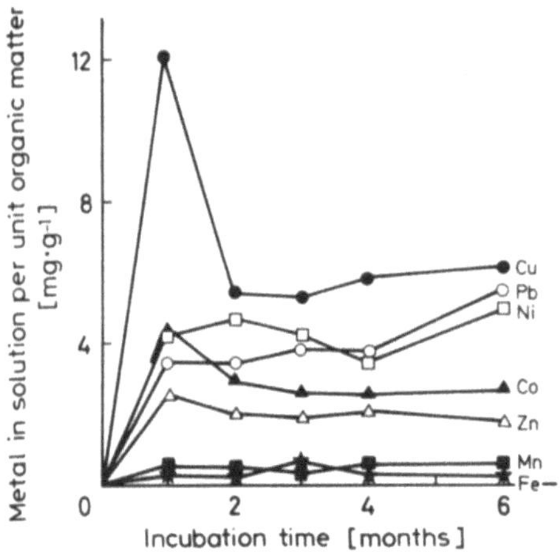

Fig. 7. Mobilization of heavy metal oxides in the presence of alfalfa residues (Data from Bloomfield et al. 1971)

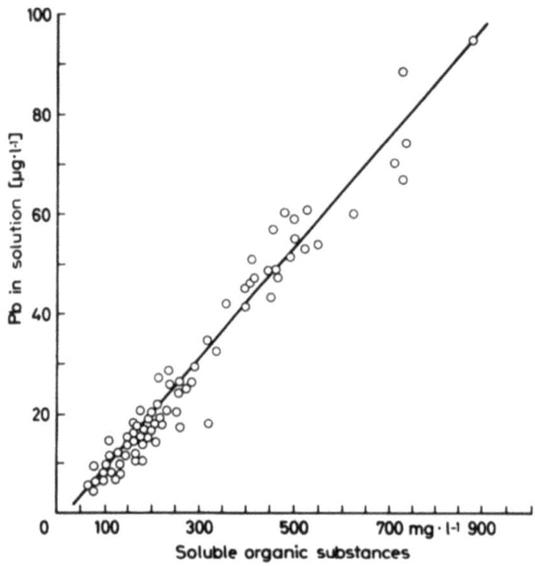

Fig. 8. Relationship between the concentration of organic substances and the concentration of Pb ions in soil percolates (Data from Tyler 1981)

The availability or non-availability of organic toxicants must be discussed in the context of "bound residues". The following conclusions can be derived from the present state of knowledge of microbial interactions with bound pesticides: (1) Living cells have the potential of accumulating organic xenobiotics, particularly lipophilic substances; (2) during the microbial degradation of organic matrices (dead microorganisms or other organic matter) residues bound to such matrices can be released into the soil solution; (3) stimulation of microbial metabolic activity (amendments, temperature, etc.) increases the release rate; and (4) bound residues are usually non-toxic residues.

3.2 Relative Toxicity of Chemicals

The potential toxicity of heavy metals can only be described in general terms. Duxbury (1985) distinguished "non-critical" (Fe only), from "potentially toxic and relatively accessible" (most accessible heavy metals have been included in previous sections) and "potentially toxic but rare or insoluble".

If compared on the basis of equal, low doses, Hg, Cd, Ag, and V display a high degree of toxicity in Duxbury's second group. Others like Zn, Cu, Mo, Co or Mn have "essential" functions (enzyme activators) at low doses in plants, animals and microorganisms and rank lower on a toxicity scale (see also Fig. 4).

The numerous organic chemicals which are used as pesticides can broadly be classified as: non-specific general biocides (e.g. fumigants), and chemicals which are more or less specific at best against a narrow range of targets (weeds, pests, pathogenic microorganisms). The overall relative toxicity of the first group is high, while that of the second group is low.

If the severity of reported side-effects (within the range of recommended doses) is used as a criterion of toxicity and if a distinction is made between reversible and apparently persistent effects on soil microorganisms (Domsch et al. 1983), the following distribution emerges (Table 2).

It can be seen from Table 2 that the frequently assumed negative impact on soil microorganisms is not necessarily linked to all pesticides but rather is restricted to a small number of cases. In the critical cases, fumigants are often involved as well as some expected effects such as herbicidal action on soil algae or fungicidal action on mycorrhizal associations (see also Table 1).

Table 2. Distribution of reported reversible and persistant effects caused by pesticides (Data from Domsch 1985)

Ecological significance	Percentage of all persistant effects	Percentage of all reversible effects
Negligible	83	36
Tolerable	12	57
Critical	5	7

3.3 Relative Stability of Chemicals

Most soil microorganisms require organic carbon for growth and maintenance. The wide diversity of catabolic reactions performed in mixed microbial communities includes the potential of also utilizing chemicals which do not normally occur in nature. Again the most prominent group of xenobiotics are pesticides. Table 3 summarizes the present status by using the half-life of chemicals as a means to describe at least the first phase of dissipation. For almost all practical considerations, a half-life of up to 100 days could be tolerated. Table 3 shows that

Table 3. Dissipation of pesticides from soils (Data from Domsch in preparation)

		Number of chemicals with half-life		
	Investigated	≤ 10 days	11–100 days	> 100 days
Herbicides	90	23	62	5
Insecticides	70	20	43	7
Fungicides	32	14	15	3

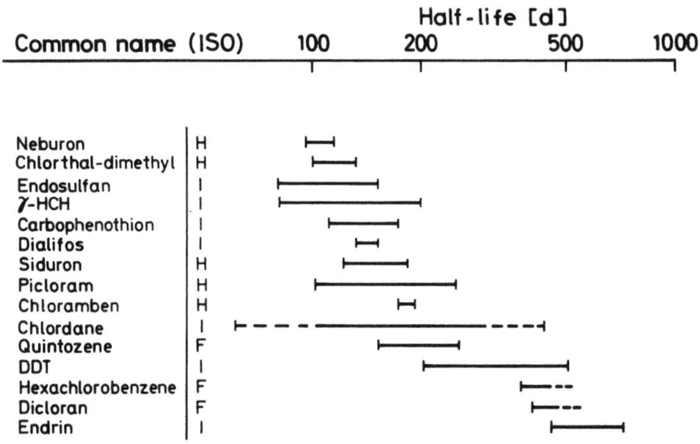

Fig. 9. Half-life ranges of persistant herbicides (**H**), insecticides (**I**) and fungicides (**F**) (Data from Domsch in preparation)

90% of the investigated pesticides belong to this group. The 15 remaining, more persistant chemicals are listed by name and half-life ranges in Fig. 9. Most of these chemicals are well known for their persistance. Some have been withdrawn from the market for that reason and none has been introduced later than 1967.

4 Environmental Influences on Chemical Behaviour

Although biotic and abiotic properties can be accurately described for each individual soil, in nature they expand over a wide range even if only the most frequently occurring soil types and textures are considered. In addition, external variables such as temperature and water regime have a modifying influence. In the following sections examples will be given to illustrates the range in which soil properties and environmental variables become effective. Emphasis will be on the degradation of pesticides as this area is sufficiently studied to allow for some generalizations.

4.1 Biotic Influences on Degradation

One of the more stable biotic properties of soils is the microbial biomass. High content of microbial biomass not only provides a high degradation potential but it also increases the probability of encounters between pesticide molecules and reactive cells. Typical arable soils contain 200–800 mg microbial carbon per kg soil. Figure 10 illustrates the linear relationship between the degradation of two herbicides and the content of microbial biomass.

Besides this quantitative aspect, the metabolic state of a microbial community has a significant influence on the course of degradation. Microbial communities in a resting state, for example, under nutrient stress, require a lag phase for reactivation and/or enzyme induction. A rapid metabolic adaptation of native microbial communities is illustrated in Fig. 11. In this case the herbicide 2,4-D serves as a suitable carbon source and within 10 days the decarboxylation potential increases within the community. Similar responses to repeated application have been described for many herbicides and insecticides.

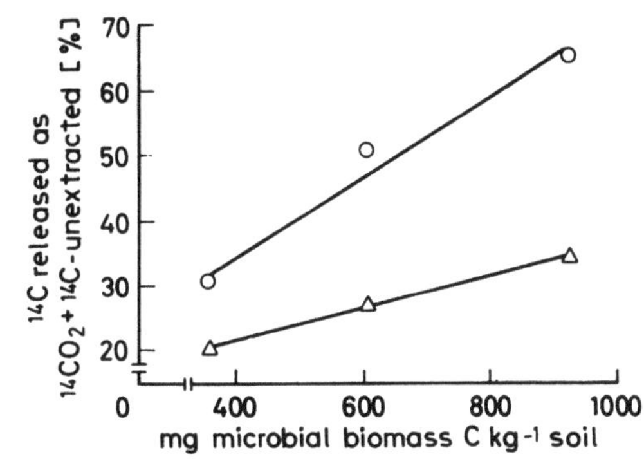

Fig. 10. Influence of quantities of microbial biomass on the degradation of carbonyl-^{14}C-labelled diallate (o—o) and allyl-2-^{14}C-labelled triallate (Δ—Δ) (Data from Anderson 1984)

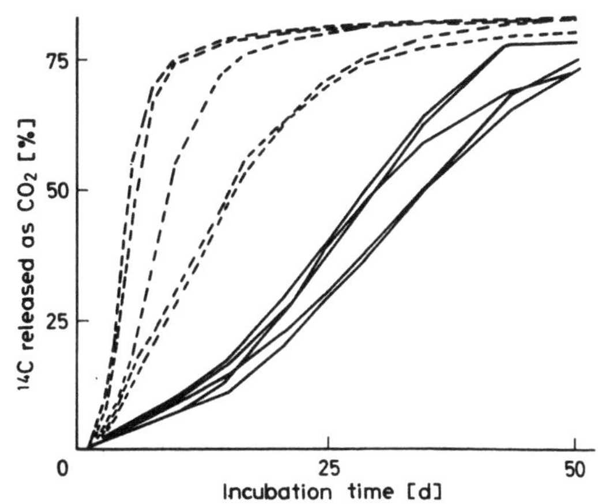

Fig. 11. Influence of repeated application on the release of $^{14}CO_2$ from carbonyl ^{14}C-labelled 2,4-D (herbicide). First application (*solid lines*), after 10 days second application (*dotted lines*) of 33 mg kg^{-1} soil. Individual curves of five parallel experiments in each group (Data from Fournier et al. 1981)

Simultaneous incubation with easily accessible, natural organic substances can result in a biomass increase and thereby accelerate degradation. The opposite can also occur, however, if the selection of a "zymogenic" community represses the production of enzymes which are required for the breaking of unusual molecule linkages.

4.2 Abiotic Influences on Degradation

The water status of soils has two aspects: total soil water content and soil water potential. The availability of water has a direct influence on microbial metabolism and usually becomes limiting at 10^2 kPa. However, the amount of water determines the effectiveness of water as a solvent and transport agent for pesticide molecules.

The soil temperature has a direct influence on the metabolic activity of the microbial biomass and a more or less direct effect on non-biological reactions of the chemical. From the examples given in Fig. 12 a $Q_{10} > 2$ can be calculated for the physiological temperature range between 5° and 25°C.

The organic matter content of soil has an immediate influence on the availability of organic molecules. In arable soils with an average organic matter content of 1–2%, the sorption capacity of this soil component does not have a large effect on rate (Fig. 13). However, histosols with 40% carbon can reduce degradation considerably. There is not a constant relationship between organic matter content and degradation because the chemicals may be adsorbed by high or low energy bonds and may be characterized by different sorption and desorption isotherms. From Fig. 13 it can also be seen that for one of the three chemicals (pentachlorophenol) a different response to decreasing contents of organic matter was obtained. A result this type would be expected in systems

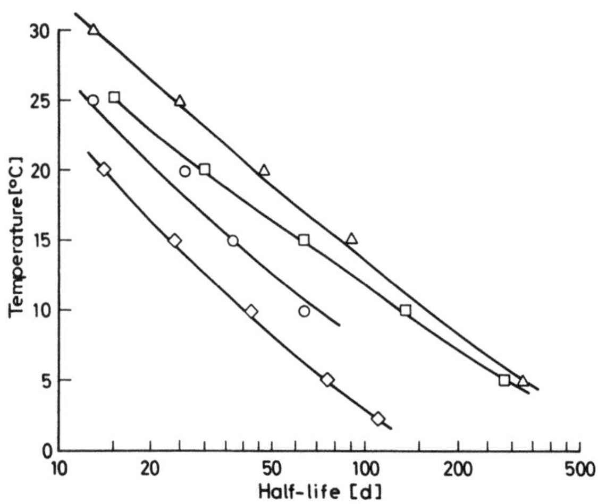

Fig. 12. Influence of temperature on the degradation of the herbicides chlorthal-dimethyl (△), propyzamide (□), chlorsulfuron (○) and the insecticide ethoprophos (◊) (Data from Domsch in preparation)

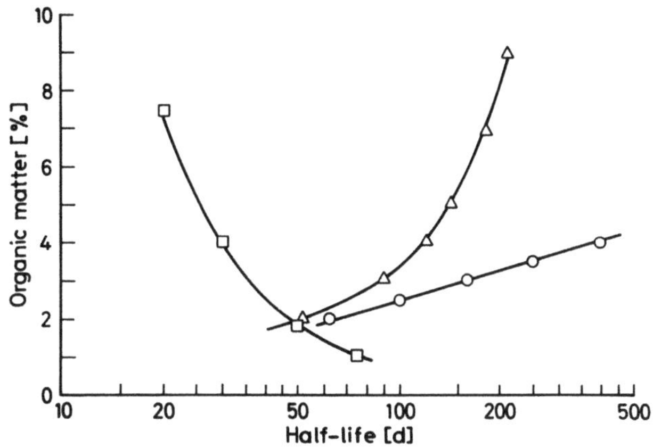

Fig. 13. Influence of organic matter on the degradation of pesticides: picloram (○), pentachlorophenol (□), dichlobenil (△) (Data from Domsch in preparation)

where microbial biomass is rate-limiting. This limitation becomes more pronounced as the amount of organic matter decreases.

Since soil pH acts on the abiotic hydrolysis and adsorption of chemicals as well as on microbial performance, there is no consistent effect of pH and therefore no generalizations can be made.

5 Conclusions

It is advisable to look at the whole microbial community if reactions of microorganisms in polluted soils are to be considered. These communities have a wide potential for adaptation, detoxification, solubilization, immobilization, transformation and mineralization. Since some of these abilities cannot be detected in short-term experiments, sufficient time must be provided in monitoring adaptation, activation and regeneration. The most critical situation from the standpoint of pollution would be one that results in persistant, irreversible chemical effects. The high sensitivity of individual bacterial or fungal species can rarely be used in biomonitoring due to the extremely time-consuming procedure involved in a reliable quantification of microbial indicator species.

It is advisable to assume that non-degradable chemicals, such as heavy metals, cannot be contained in an inactive state and consequently are a potential hazard to all forms of life in soil. There is good reason to minimize application of heavy metal-containing "amendments" to all soils. Degradable chemicals such as pesticides will not cause irreversible effects provided frequency and rates of application do not increase and thus eventually exceed the assimilatory capacity of the soil.

The "soil aspect" requires the consideration of physical, chemical and biological properties in order to recognize or predict the behaviour of potentially toxic chemicals. Table 4 lists four properties (organic matter, carbonate and clay content and pH) represented in a selection of four frequently occurring soil types. For practical and theoretical reasons, these soils have been recommended for laboratory or field testing of chemicals. It is advisable to extend the characterization of these soils to the content of microbial biomass whenever microbial communities or functions are part of the complex chemical behaviour.

Table 4. Recommended test soils (FRG) concerning sorption/desorption, percolation, side-effects, degradation (Data from Führ et al. 1983)

	Organic matter (%)	$CaCO_3$ (%)	Clay (%)	pH ($CaCl_2$)
Phaeozem-, chernozem- or calcaric fluvisol-Ap	3.5–4.2	0.5–1.5	13–17	7.2–7.5
Dystric cambisol-Ap	1.3–1.8	–	6–10	5.5–6.0
Podsol-Aeh	7–9	–	2	3.5
Dystric regosol-B or dystric gleysol-G	0.2	–	50–60	4–5

References

Al-Khafaji AA, Tabatabai MA (1979) Effect of trace elements on arylsulfatase activity in soils. Soil Sci 127:129–133

Anderson JPE (1984) Herbicide degradation in soil: influence of microbial biomass. Soil Biol Biochem 16:483–489

Barkay T, Tripp SC, Olson BH (1985) Effect of metal-rich sewage sludge application on the bacterial communities of grasslands. Appl Environ Microbiol 49:333–337

Beck Th (1981) Untersuchungen über die toxische Wirkung der in Siedlungsabfällen häufigen Schwermetalle auf die Bodenmikroflora. Z Pflanzenernähr Bodenkd 144:613–627

Bloomfield K, Kelso WI, Piotrowska M (1971) The mobilisation of trace elements by aerobically decomposing plant material under simulated soil conditions. Chem Ind 59–61

Brookes PC, McGrath SP, Klein DA, Elliott ET (1984) Effects of heavy metals on microbial activity and biomass in field soils treated with sewage sludge. Environ Contam Program Edinburgh, pp 574–583

Cole MA (1977) Lead inhibition of enzyme synthesis in soil. Appl Environ Microbiol 33:262–268

Coughtrey PJ, Jones CH, Martin MH, Shales SW (1979) Litter accumulation in woodlands contaminated by Pb, Zn, Cd and Cu. Oecologia 39:51–60

Domsch KH (1985) Funktionen und Belastbarkeit des Bodens aus der Sicht der Bodenmikrobiologie. Materialien zur Umweltforschung 13:1–65. Kohlhammer, Stuttgart Mainz

Domsch KH Pestizide und Bodenmikroorganismen. (in preparation)

Domsch KH, Jagnow G, Anderson TH (1983) An ecological concept for the assessment of side-effects of agrochemicals on soil microorganisms. Residue Rev 86:65–105

Duxbury T (1985) Ecological aspects of heavy metal responses in microorganisms. Adv Microbiol Ecol 8:185–235

Fournier JC, Codaccioni P, Soulas G (1981) Soil adaptation to 2,4-D degradation in relation to the application rates and the metabolic behaviour of the degrading microflora. Chemosphere 10:977–984
Führ F, Biehl HM, Thielert W (1983) Methoden zur ökologischen Bewertung von Chemikalien, vol 2: Böden und Modellsysteme. Jül-Spez 224:1–107
Gildon A, Tinker PB (1983) Interactions of vesicular-arbuscular mycorrhizal infection and heavy metals in plants. I. The effects of heavy metals on the development of vesicular-arbuscular mycorrhizas. New Phytol 95:247–261
Juma NG, Tabatabai MA (1977) Effects of trace elements on phosphatase activity in soils. Soil Sci Soc Am J 41:343–346
Kanazawa S, Filip Z (1986) Effects of trichloroethylene, tetrachloroethylene and dichloromethane on enzymatic activities in soil. Appl Microbiol Biotechnol 25:76–81
Mathur SP, Hamilton HA, Lévesque MP (1979) The mitigating effect of residual fertilizer copper on the decomposition of an organic soil in situ. Soil Sci Soc Am J 43:200–203
Mathur SP, MacDougall JI, McGrath M (1980) Levels of activities of some carbohydrases, protease, lipase and phosphatase in organic soils of different copper content. Soil Sci 129:376–385
Nordgren A, Bååth E, Söderström B (1983) Microfungi and microbial activity along a heavy metal gradient. Appl Environ Microbiol 45:1829–1837
Rühling A, Bååth E, Nordgren A, Söderström B (1983) Fungi in metal-contaminated soil near the Gusum Brass Mill, Sweden. Ambio 13:34–36
Tabatabai MA (1977) Effects of trace elements on urease activity in soils. Soil Biol Biochem 9:9–13
Trevors JT, Stratton GW, Gadd GM (1986) Cadmium transport, resistance, and toxicity in bacteria, algae, and fungi. Can J Microbiol 32:447–464
Tyler G (1976) Heavy metal pollution, phosphatase activity, and mineralization of organic phosphorus in forest soils. Soil Biol Biochem 8:327–332
Tyler G (1981) Heavy metal in soil biology and biochemistry. Soil Biochem 5:371–414
Zamani B, Knezek BD, Dazzo FB (1984) Biological immobilization of zinc and manganese in soil. J Environ Qual 13:269–273

Part C Management of Inorganic Pollutants

Introductory Comments. Proper and efficient solid waste disposal presents a difficult problem to government regulating bodies, design engineers, agronomists, public health officials and the general public. Solid waste from almost all sources presents a dual face: it is beneficial to agriculture and at the same time it contains toxic elements, mainly heavy metals. Regardless of its properties we keep producing solid waste in large quantities and a proper way must be established to dispose of all kinds of waste.

The three chapters in Part C address two aspects of solid waste disposal: sewage sludge and utility ashes. Table 1 lists maximum permissible concentrations of heavy metals in sewage sludge applied to agricultural fields in different European countries. The figures show that (1) there are large differences in the permitted concentrations among the various countries; and (2) the degree of removal required, and hence the cost of sewage treatment, may vary considerably. The removal required in three different locations in the USA and England can be estimated by comparing realistic heavy metal concentrations in sewage sludges as reported herein by Chang et al., Beckett and Chaney, with the EEC permissible concentrations presented in Table 1.

Table 1. Maximum permissible heavy metal concentration in sewage sludge (mg kg^{-1} dry solids) applied to agricultural fields in different countries[a]

	CH, D, F[b]	DK	EEC	SW	IL
Hg	10	6	16	5	n.s.
Cd	30	8	30 ± 10	4	8 ± 3
Ni	200	40 ± 20	350 ± 50	100	42
Cr	100 ± 100	80 ± 40	750	130	90 ± 10
Pb	1000 ± 200	475	880 ± 120	200	60 ± 40
Cu	100 ± 100	500	1250 ± 200	600	150 ± 50
Zn	3000	3000 ± 1000	2750 ± 250	1500	1500 ± 500

[a] Malz, F. (1986) Heavy metals and chlorinated hydrocarbon in sewage sludge. Proceedings of IUPAC Workshop, Frankfurt 15–16 October 1986.
[b] Country names abbreviations (left to right): CH, Switzerland; D, West Germany; F, France; DK, Denmark; EEC, European Common Market; SW, Sweden; IL, Israel.

Chang et al. studied the beneficial effects that sewage sludge and utility ash application have on the physical properties of soil, as well as the penalty paid. Chaney, on the other hand, discussed various approaches to estimate the potential risks involved in applying different sludges to soils differing in chemical properties, and to crops which differ in heavy metal uptake efficiency and rate of ingestion by the population. Unlike Chaney's optimistic conclusion that sludge utilization will become safer and of higher benefit to agriculture, Beckett concluded that current knowledge does not allow one to estimate heavy metal accumulation in edible crops grown on sludge-amended soils. The difficulties in estimating such accumulation arise from the complex route which the toxic elements take until they are spread as sewage sludge on agricultural land.

J. Goldshmid

Municipal Sludges and Utility Ashes in California and Their Effects on Soils

A.C. Chang, A.L. Page, L.J. Lund, J.E. Warneke, and C.O. Nelson[1]

1 Introduction

The worldwide economic prosperity of the post-World War II era has resulted in dramatic rises in material consumption and, consequently, solid waste generation. This pattern is especially apparent for industrialized nations of western Europe and northern America. In the USA, an estimated 2.6 billion tons of solid wastes are produced annually. All of them pose threats of environmental pollution and require proper disposal. Compared to wastes of mining, agricultural, and industrial origins, municipal sludges and utility ashes are far less significant in quantities. Because they are frequently applied on land, their impact on the soil environment must be evaluated.

Municipal sludges are residues resulting from the treatment of sewages and municipal wastewater. Typically, 50–90% of the solids in sewage sludges are organic. As the conglomerate of impurities removed from contaminated water, they always have a high concentration of environmental pollutants. At the time sludges are produced, they contain biodegradable organic matter, pathogens, toxic chemicals, essential plant nutrients, and dissolved mineral solids. In land application, any one of them can become the limiting factor. Unlike the sludges, ashes from the combustion of coal for power generation are entirely inorganic and they consist primarily of iron and aluminum silicates with an array of other chemical elements in lesser quantities. Their chemical compositions are influenced by the geochemical nature of the coal and the method used for air pollution control. Solid wastes such as municipal sludges and coal ashes may be used to alter the agronomical properties of soils (Epstein et al. 1976; Chang et al. 1977, 1983). The amounts of waste solids required to significantly change the soil's physical properties, however, were large. Generally, 25–35% of wastes by volume must be added to accomplish the desirable effects. This practice may introduce large amounts of potentially hazardous inorganic chemical constituents into the soils.

In this report, we survey inorganic chemical properties of sludges and fly ashes in California, demonstrate movements of selected inorganic chemical constituents in the soil profile, and review the effect of solid wastes on the physical properties of soils.

[1]Department of Soil and Environmental Sciences Soils-84, University of California, Riverside, California 92521-0424 USA

2 Inorganic Chemical Composition of Fly Ash and Municipal Sludge in California

2.1 Fly Ash

When the pulverized coal is combusted in suspended burners, over 80% of the combustion residues may be fly ashes. Once formed, they are carried by the exiting flue gases and most of them will be captured by the air pollution control devices. Fly ashes collected in this manner are finely divided and spherical-shaped amorphous material with a particle texture resembling a sandy loam soil. Studies also indicated that fly ashes recovered from coal burning consist of 70–90% glass (Watt and Thorne 1965; Davison et al. 1974). Chemically, fly ash contains primarily silicates of iron and aluminum with lesser quantities of calcium, magnesium, sodium, and sulfur in oxide forms (Hecht and Duvall 1975). To understand its chemical nature, the fly ash of a coal-fired power plant in California was sampled. The collected ash samples were size classified and the elemental contents of each size fraction were assayed using neutron activation (Tables 1 and 2).

The fly ash examined was made up predominantly of Al, Si, Ca, Fe, Mg, Na, K, S, Ba, and Sr (Table 1). These elements constituted almost 50% of the fly ash and their concentrations in the size-classified fractions were not significantly different from one another. Fly ashes derived from lignite and subbituminous coal of the western USA were considerably higher in Ca, Si, and Fe than those derived from bituminous and anthracite coal of the eastern USA (Swanson et al. 1976). Intrusions of calcium carbonate and gypsum into the coal deposits were believed to be the primary cause. Although the total concentrations of major elements in the fly ash we examined were comparable to those of many southern California soils, its land application increased in the soluble Ca, Sr, and S of the soils (Phung et al. 1976; Elseewi et al. 1978). Aluminum in the fly ash, however,

Table 1. Selected major elements of a fly ash in California

Element	Unclassified	Size classified			
		< 53 m	53–150 m	150–250 m	> 250 m
	(%)				
Na	1.2	1.3	1.2	1.2	0.9
K	0.5	0.5	0.5	0.5	0.5
Ca	10.3	11.5	11.4	10.6	8.1
Fe	3.3	3.4	3.6	3.4	2.5
Al[a]	18.0	–	–	–	–
Si[a]	10.0	–	–	–	–
Mg[a]	1.7	–	–	–	–
S[a]	0.4	–	–	–	–

[a]Contents of these elements were determined spectrographically in a Na_2CO_3 fusion extract (Bradford et al. 1976); they were not size classified.

Table 2. Minor and trace elements of a fly ash in California

Element	Unclassified	Size classified			
		< 53 m	53–150 μm	150–250 μm	> 250 μm
	(μg g⁻¹)		(μg g⁻¹)		
As	14.14	17.05	13.25	12.03	10.38
Ba	3685	4089	4019	3670	2657
Ce	108	112.5	113.8	137.7	84.24
Co	13.42	14.95	14.26	13.58	10.11
Cr	64.22	64.83	71.03	65.23	51.70
Cs	4.92	4.63	5.12	5.32	4.12
Eu	1.25	1.24	1.18	1.20	0.93
Ga	28.67	34.01	32.09	35.15	27.18
Hf	7.93	8.13	8.40	8.03	6.00
La	59.5	63.52	63.26	77.68	46.51
Lu	1.24	1.31	1.33	1.24	0.88
Mo	8.77	12.62	12.13	10.19	6.96
Nd	45.07	48.00	43.16	54.80	34.88
Ni	50.35	46.31	–	37.98	29.33
Rb	53.10	51.26	53.02	55.95	44.38
Sb	3.82	4.38	3.90	3.67	2.92
Sc	16.11	17.12	17.23	16.52	12.19
Se	18.83	21.94	19.62	17.32	15.50
Sm	8.93	9.20	9.23	9.79	6.79
Sr	1839	2011	1927	1875	1313
Ta	1.50	1.58	1.50	1.70	1.21
Tb	1.02	1.15	1.30	1.04	0.81
Th	20.75	20.92	21.13	22.33	16.32
U	7.36	7.73	7.44	8.14	6.29
W	2.66	3.66	3.07	1.42	–
Yb	4.04	4.48	4.56	3.67	3.06
Zn	98.77	136.9	102.9	96.88	75.70
Zr	183.4	187.4	224.7	180.1	139.8

was tied up in an insoluble alumino-silicate matrix. Despite its high concentration in the fly ash, the solubility of this element in fly ash-soil mixtures was limited.

Concentrations of trace elements in the fly ash were considerably higher than those in the coal (Table 2). Besides B, Mo, and Se, concentrations of other trace elements in the fly ash were similar to and well within the range of those in soils. The availability of fly ash-derived B, Mo, and Se has been demonstrated by Doran and Martens (1972), Plank and Martens (1974), and Elseewi et al. (1976). Plants grown on fly ash-amended soils were enriched in these trace elements. Except for Cr, Cs, Rb, and U, the trace elements in Table 2 showed the trend of increasing concentration as the particle became finer. Effects of particle size on elemental concentrations of fly ash have been observed by many investigators (Lee and Von Lehmden 1973; Davison et al. 1974; Lee et al. 1975; and Kaakinen et al. 1975). Natusch et al. (1975) demonstrated that some elements (As, Cd, Cu, Ca, Mo, Pb, S, Sb, Se, Ti, and Zn) in fly ashes tended to increase in concentration with decrease in particle size and concentrations of other elements (Al, Ba, Ca, Ce, Co, Eu, Fe,

Hf, La, Mg, Mn, Rb, Sc, Si, Sm, Sr, Ta, Th, and Ti) in fly ashes were not affected by particle size. There were also elements (Be, Cr, K, Na, Ni, Sc, U, and V) whose concentration distribution patterns in size-classified fly ashes were in between the above two groups. The mechanism by which small particles of fly ash were enriched with trace elements was primarily a surface phenomenon. Elements which volatilized, partially or completely, at the temperature of combustion would condense at the surface of fly ash particles as the temperature of the flue gases dropped. Because of their large specific areas, smaller fly ash particles resulted in higher trace element concentrations. Trace elements that are deposited on the surface of fly ash particles have greater opportunities to react chemically in the environment.

2.2 Municipal Sewage Sludges

Land application is by far the most common method of disposing sewage sludges in California. To assess the environmental impact of land applications, chemical compositions of sewage sludges must be well understood. A survey was conducted to evaluate the inorganic chemical composition of municipal sludges in California. Over a 4-year period, sludge samples were obtained from 110 publicly owned treatment works which serviced 75% of the state's sewered population. These facilities represented communities of varying population sizes, economic conditions, and geographical locations. Collected sludge samples were dried at 65 °C and ground to pass a 1-mm sieve. They were analyzed for total N, P, K, S, Al, Fe, Cd, Cr, Cu, Ni, Pb, and Zn. Tables 3 and 4 summarize the results of the study.

Contents of selected essential plant nutrients and Al in sludges were tabulated according to types of treatment plants (Table 3). Within each type of treatment, N, P, K, and S concentrations of sludges followed a normal distribution. Unusually high concentrations of Al and Fe in some of the sludges reflected the uses of Al and Fe in sludge dewatering. The concentrations of K, S, Fe, and Al in sludges did not appear to be affected by the type of treatment plant.

Table 3. Fertilizer elements of municipal sludges in California

Element	Primary anaerobic Range	Mean	Trickling filter anaerobic Range	Mean	Waste-activated anaerobic Range	Mean	Waste-activated aerobic Range	Mean
	%							
N	1.20–3.28	2.20	2.41–4.09	3.18	3.02–5.01	4.02	4.30–7.06	5.78
P	0.30–2.25	1.14	1.10–2.67	1.64	1.27–2.90	2.03	1.42–4.75	2.43
K	0.12–0.55	0.29	0.10–0.60	0.25	0.10–0.50	0.26	0.12–0.72	0.36
S	0.31–2.90	1.01	0.45–2.26	1.05	0.43–1.71	1.04	0.56–0.98	0.78
Fe	0.90–7.95	2.12	0.60–2.80	1.44	0.75–6.50	1.55	0.45–1.70	0.93
Al	0.85–4.95	1.88	0.75–4.20	1.55	0.85–4.80	1.56	0.45–2.05	1.04

Table 4. Heavy metals in California sludges

Element	Range	Median	Mean	Weighted average
		($\mu g\ g^{-1}$)		
Cd	5-315	11	28	89
Cr	23-21000	161	602	1269
Cu	217-4100	630	776	1302
Ni	20-2300	50	138	325
Pb	91-9800	504	726	821
Zn	665-5862	1680	1838	2730

On the other hand, the concentrations of N and P were low in primary sludges and high in waste-activated sludges.

Cadmium, Cr, Cu, Ni, Pb, and Zn are commonly used industrial metals and their concentrations in the sludges of different communities varied by orders of magnitude (Table 4). The distribution of these metals in sludges was log-normal. Sludges with metal contents at the low end of the distribution represented wastewater primarily of residential origins. A few sludges which were very high in metals reflected significant contributions from industrial discharges (Klein et al. 1974). The range of concentrations agreed well with those obtained elsewhere in the USA and in other countries (Blakeslee 1973, Sommers 1977; Tabatabai and Frankenberger 1979; Beckett 1980).

3 Movement of Toxic Chemical Constituents in Solid Waste-Treated Soils

Biochemically toxic inorganic constituents in solid wastes are often present in the form of cations. Nieboer and Richardson (1980) proposed that elements of environmental and toxicological importance be classified in a biologically and chemically significant manner. Their classification was derived from the work of Ahrland et al. (1958) who classified cations into three groups according to trends in the equilibrium constants of cation/ligand complexes. They are:

Class A: Li^+, Na^+, K^+, Cs^+, Ca^{2+}, Ba^{2+}, Sr^{2+}, Sr^{2+}, Mg^{2+}, Be^{2+}, La^{3+}, Y^{3+}, Gd^{3+}, Lu^{3+}, Sc^{3+}, and Al^{3+}.
Class B: Ag^+, Au^+, Tl^+, Cu^+, Hg^{2+}, Pd^{2+}, Bi^{2+}, Tl^{2+}, and Pb^{2+}.
Borderline: Mn^{2+}, V^{2+}, Ti^{2+}, Zn^{2+}, Cr^{2+}, Ni^{2+}, Fe^{2+}, Co^{2+}, Cd^{2+}, Cu^{2+}, Sn^{2+}, Fe^{3+}, Ga^{3+}, As^{3+}, Sn^{3+}, Pb^{4+}, In^{4+}, and Sb^{4+}.

Generally, Class B ions are expected to be more toxic than borderline ions, which, in turn, are more toxic than Class A ions. Their toxicities are the results of chemical and/or biochemical mechanisms that block the essential functional groups of biomolecules, displace essential metal ions in biomolecules, and modify the active conformation of biomolecules (Ochiai 1977). The Class B ions and, to a lesser extent the borderline ions, are chemically more likely to take part in reactions resulting in biotoxicities. According to this classification, many chemical elements found in the fly ash and the municipal sludges may be potentially toxic (Table 1–4).

To examine the fate of inorganic constituents in solid waste-treated soils, the movements of Na (a Class A ion), Cd (a borderline ion), and Pb (a Class B ion), in a long-term sludge land application site were used as examples (Tables 5 and 6, Fig. 2). From 1975 to 1983, composted sludges from the Joint Water Pollution Control Plant of the County Sanitation Districts of Los Angeles County were applied on land at rates of 0, 22.5, 45, and 90 metric tons per hectare per year. The Na ion introduced into the soil by the waste applications and irrigations slowly moved downward in the soil profile (Fig. 2). In the same period, the Cd and Pb added remained in the surface soils (Tables 5–6). While the concentrations of Cd in the surface 30 cm of the soil profile continued to rise with sludge applications, no increase of soil Cd and Pb was detected in the lower strata. The lack of mobility of borderline and Class B metal ions such as the Cd and Pb indicated strong chemical reactions in the soil to attenuate the toxicological effects of metal ions. Even at very high application rates, toxic effects of metal ions on plants were rarely

Table 5. 4 N HNO_3 extractable Cd in composted sludge-treated Domino loam soil

Sludge treatment	Soil depth	Year[a]			
		1975	1978	1981	1986
	(cm)	(μg g⁻¹ dry wt.)			
	0–15	0.8	.7	0.5	0.4
0	15–30	0.5	0.6	0.2	0.1
(Control)	30–45	0.7	0.7	0.1	[b]
	45–60	0.7	0.6	0.1	[b]
	0–15	0.7	1.3	3.7	4.0
22.5	15–30	0.5	0.6	0.4	0.7
	30–45	0.6	0.6	0.1	[b]
	45–60	0.6	0.6	0.1	[b]
	0–15	0.8	2.1	6.0	6.2
45	15–30	0.8	0.8	0.5	0.2
	30–45	0.8	0.8	0.2	[b]
	45–60	0.8	0.8	0.1	[b]
	0–15	0.7	0.1	11.9	14.4
90	15–30	0.5	0.8	0.8	1.4
	30–45	0.6	0.7	0.2	[b]
	45–60	0.6	0.7	0.1	[b]

[a]Values for 1975 represented metal levels prior to sludge application; sludge application ceased in 1983.

[b]Soil samples were not obtained.

Table 6. 4 N HNO_3 extractable Pb in composted sludge-treated Domino loam soil

Sludge treatment	Soil depth	Year[a] 1975	1978	1981	1986
	(cm)	($\mu g\ g^{-1}$ dry wt.)			
	0–15	17	13	14	16
0	15–30	25	10	8	10
(Control)	30–45	21	11	7	[b]
	45–60	20	10	8	[b]
	0–15	13	37	76	78
22.5	15–30	23	11	14	26
	30–45	18	10	7	[b]
	45–60	18	11	7	[b]
	0–15	16	65	119	110
45	15–30	23	12	17	11
	30–45	22	11	7	[b]
	45–60	20	11	8	[b]
	0–15	15	114	237	212
90	15–30	15	15	23	48
	30–45	17	14	8	[b]
	45–60	18	13	7	[b]

[a]Values for 1975 represented metal levels prior to sludge application; sludge application ceased in 1983.
[b]Soil samples were not obtained.

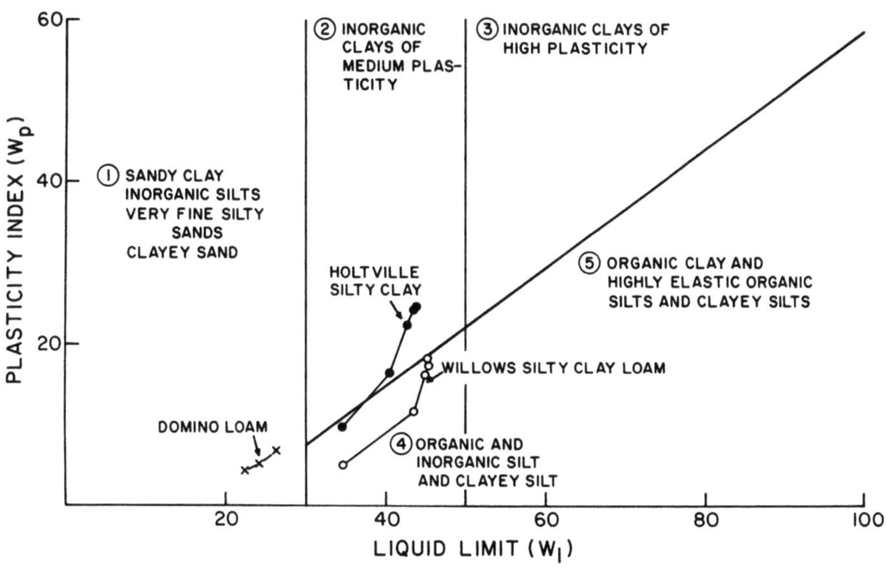

Fig. 1. Classification of fly ash-soil mixture by Casagrande Plasticity Chart

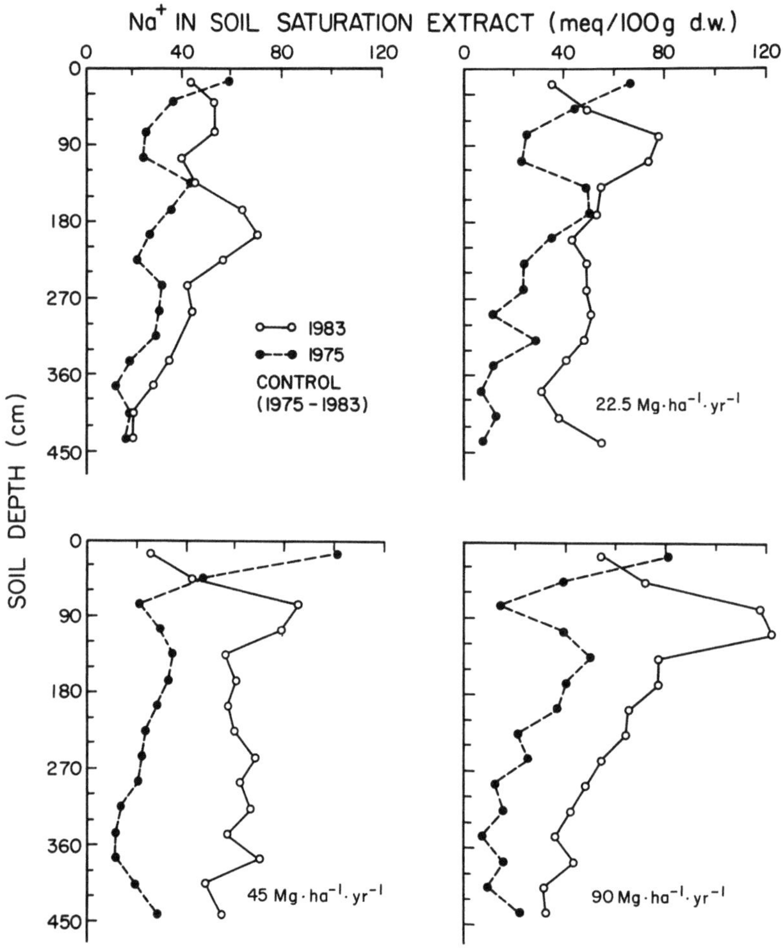

Fig. 2. Distribution of Na in the soil saturation extract of composted sludge-treated Domino soils

reported (Logan and Chaney 1983). It, however, did not prevent the vegetation from accumulating sufficient toxic metals to be a potential hazard to the consumers (Chaney 1987).

4 Physical Properties of Waste-Treated Soils

When solid wastes are incorporated into the soil, they may alter the physical properties of the soils. The physical properties of a soil are frequently characterized by measurement of the soil's particle size distribution, specific gravity, soil water characteristics, and consistency limits. The fundamental characteristics of the soils would, in turn, influence the soil's mechanical and physical behavior. For agronomic or engineering uses of the soil, it is essential that the physical behavior of a soil is fully understood.

To examine the effects of solid wastes on the physical properties of soils, three southern California soils were selected for experiments. They were Holtville silty clay (clay over loam, montmorillonitic [calcareous], hyperthermic typic Torrifluvent), Willows silty clay loam (fine, montmorillonitic, thermic typic Pelloxerert), and Domino loam (fine-loamy, mixed, thermic xerollic Calciorthid). A composted sewage sludge and a fly ash were mixed, in incremental amounts, with the soils and equilibrated at 0.2 bar soil moisture tension for 48 h. The equilibrated waste-soil mixtures were then packed into brass rings, 5 cm in diameter and 5 cm in height, for measurements of bulk density, water-holding capacity ca. 0.2 bar, saturated hydraulic conductivity, and modulus of rupture. Details of the measuring procedures may be found in Richards et al. (1965) and Chang et al. (1977 and 1983). In addition, the particle size distribution and consistency limits of fly ash-treated soils were determined (Bouyoucos 1951; Day 1965; and ASTM 1970). To simulate effects of weathering on the physical properties of fly ash-treated soils, the soil-fly ash mixtures also were subjected to 12 weekly wetting and drying cycles prior to the measurements.

4.1 Particle Size Analysis and Consistency Limits

In general, the incorporation of solid wastes affected the distribution of particles in the soils (Table 7). The changes in the distribution were directly related to the initial distribution of particles and the proportions of soil and wastes in the mixture. As the fly ash used in this experiment consisted of 37% sand, 52% silt, and 11% clay, adding fly ash to the heavy textured soil (Holtville silty clay) reduced the clay fraction and proportionally increased the sand and silt fractions. At the 50% fly ash incorporation, the clayey soil was transformed into a loamy mixture.

The textural changes of the soils induced by adding solid wastes were confirmed by measurements in the consistency limits (Table 8). The consistency limits define the range of moisture contents at which a soil would behave as a solid

Table 7. Particle size distribution of fly ash-treated soils

Fly ash added	Holtville silty clay			Willows silty clay loam			Domino Loam		
	Sand	Silt	Clay	Sand	Silt	Clay	Sand	Silt	Clay
(%)		(%)			(%)			(%)	
0	12	41	47	8	54	38	52	28	20
6.25	13.6	41.7	44.7	9.8	53.8	36.3	31.1	29.5	19.4
12.5	15.1	42.4	42.5	11.6	53.8	34.6	50.1	31.0	18.9
25	18.3	43.8	38.0	15.3	53.5	31.3	48.3	34	17.8
50	24.5	46.5	29	22.5	53	24.5	44.5	40	15.5
100	37	52	11	37	52	11	37	52	11

Table 8. Consistency limits of fly ash-affected soils

Soil type	Fly ash	Liquid limit		Plastic limit		Shrinkage limit	
		Non-weathered	Weathered[a]	Non-weathered	Weathered	Non-weathered	Weathered
	(%)	(%)					
Holtville clay	0	45[b]	45 a	21 b	19 a	14 c	14 c
	6.25	48 a	42 b	21 b	19 a	16 c	17 bc
	12.5	46 ab	40 bc	23 b	19 a	22 b	19 ab
	25	43 b	38 c	29 a	20 a	26 ab	22 a
	50	36 c	34 d	28 a	21 a	27 a	22 a
Willow silty clay	0	46 a	46 a	29 b	33 b	29 bc	28 a
	6.25	46 a	46 a	31 b	34 ab	27 a	28 a
	12.5	46 a	44 ab	32 b	33 b	29 bc	28 a
	25	46 a	41 b	36 a	34 ab	33 a	30 a
	50	36 b	34 c	38 a	37 a	31 ab	28 a
Domino silty loam	0	26 a	27 a	24 a	23 b	17 a	21 a
	6.25	26 a	23 b	24 a	26 ab	18 a	19 a
	12.5	25 a	25 ab	27 a	28 a	18 a	22 a
	25[c]	—	—	—	—	—	—
	50[c]	—	—	—	—	—	—

[a] Weathered and nonweathered treatments denote fly ash-soil mixtures were and were not subject to wetting and drying prior to measurements.
[b] Within each combination, parameter values followed by the same letter indicate they are not significantly different at $P < 0.05$ by Duncan multiple range test.
[c] Fly ash-soil mixture became noncohesive.

(shrinkage limit), semi-solid (plastic limit), or liquid (liquid limit). The effects due to fly ash treatments on soil properties were statistically tested using one-way analysis of variance at $P < 0.05$. For the coarsely textured soil (Domino loam), adding small amounts of fly ash did not affect its consistency limits. As the fly ash addition exceeded 25%, the Domino soil, however, completely lost its plasticity. The soil moisture content required one to keep the fly ash-soil mixtures of the heavily textured soils (Holtville silty clay and Willows silty clay loam) in a liquid state (the liquid limit) decreased with increasing amounts of fly ash used. In the meantime, the minimum moisture content required to maintain the fly ash-soil mixtures in the plastic state (plastic limit) increased with the amounts of fly ash added. As the result, the plastic index, whose numerical value was the difference between the liquid limit and the plastic limit, of fly ash-treated Holtville and Willows soils became smaller as the amounts of fly ash in soils increased. The narrowing of the moisture range in which a soil would remain in the plastic state indicated the lessening of the soil's cohesiveness due to the presence of fly ash. The change of the soil's consistency properties might also be illustrated by plotting them on a Casagrande plastic chart (Fig. 1) (Casagrande 1948). As amounts of fly ash in the soil increased, the mechanical characteristics of the fly ash-soil mixtures gradually shifted from a claylike material (cohesive) to a siltylike material (less cohesive).

4.2 Bulk Density, Hydraulic Conductivity, Water-Holding Capacity, and Modulus of Rupture

Fundamental changes in the soil texture caused by waste amendment also affected other physical properties of the solid waste-treated soils. Bulk density, hydraulic conductivity, water-holding capacity, and modulus of rupture may be used to characterize the physical behavior of soils used for crop production. As fly ash and sewage sludge were introduced, the soils experienced modifications in their physical behavior (Tables 9 and 10).

Depending on the densities of the wastes and those of soils, additions of fly ash and sewage sludge either reduced or increased the bulk densities of the waste-soil mixtures. The magnitude of changes, however, were not directly proportional to the amount of wastes added. Generally, small amounts of amendment did not affect bulk densities of the waste-soil mixtures. The percentage of solid wastes in the mixtures had to reach 25% and sometime 50% before significant changes in bulk density would take place. The water-holding capacity (measured as volumetric water content at 0.2 atm) of all waste-treated soils increased and the pattern of changes was similar to those of their bulk densities. When small amounts of fly ash were added into the soils (< 12.5%), the hydraulic conductivity of soils increased. The effect of fly ash on the hydraulic conductivity of soils was probably related to the ability of the fly ash to coagulate and flocculate soil particles (Goecker et al. 1956). For fly ash-soil mixtures that were not subject to wetting and drying, their hydraulic conductivities remained high with addi-

Table 9. Physical properties of fly ash and composted sludge-treated Domino loam

Amount added (%)	Bulk density ($g\ cm^{-3}$)	Volumetric water content ca. 0.2 atm ($g\ g^{-1}$)	Hydraulic conductivity ($cm\ h^{-1}$)	Modulus of rupture ($kg\ cm^{-2}$)
	(a) Fly ash			
0	1.37 a[a]	0.22 b	0.2 d	2.22 a
6.25	1.30 ab	0.22 b	4.2 ab	0.23 b
12.5	1.27 ab	0.22 b	5.2 a	0.23 b
25	1.24 b	0.22 b	3.4 b	0.06 c
50	1.18 b	0.28 a	1.1 c	0.06 c
	(b) Composted sludges[b]			
0	1.45 a	0.23 b	<0.1 a	2.63 a
6.25	–	–	–	–
12.5	1.37 a	0.25 b	<0.1 a	1.47 b
25	1.31 a	0.27 b	0.1 a	0.85 c
50	1.04 b	0.35 a	2.6 b	0.47 c

[a] Within each data group, values followed by the same letter indicate they are not significantly different at $P < 0.05$ by Duncan multiple range test.
[b] From Table 1 in Chang et al. (1983).

tions of fly ashes (up to 25%). However, the improvement in hydraulic conductivity dissipated as fly ash addition reached 50%. The enhancement of hydraulic conductivities of the fly ash-treated soils was not observed when the mixtures were subjected to cycles of wetting and drying prior to the measurement (Table 10). The effect of organic solids on the ability of soil to conduct water appeared to be longer-lasting. When adequate amounts of sludges were added, hydraulic conductivity of the soils could be improved. Even years after sludge incorporations, improved hydraulic conductivity of sludge-treated soils could still be measured in the field (Chang et al. 1983). The modulus of rupture of a soil measured the shear stress which reflected the magnitude of the interlocking force and adsorption force between small soil particles (Reeves 1969). Both wastes, even incorporated in small quantities, reduced the modulus of rupture of waste-soil mixtures (Table 9). As the result, the soil's ability to bear stress would be reduced (Amos and Wright 1972).

In a solid waste-treated soil, the modification of the soil's physical behavior is due primarily to the addition of bulk solids which alter the particle size distribution and composition. Constituents added with the waste may also result in chemical reactions that change soil structure and soil aggregation. The role of waste application-induced chemical reactions on the physical properties of soils, however, is unknown.

Table 10. Physical properties of weathered and unweathered fly ash-treated soils

Material	Fly ash added	Bulk density Non-weathered	Bulk density Weathered	Volumetric water content at 0.2 atm Non-weathered	Volumetric water content at 0.2 atm Weathered	Saturated hydraulic condition Non-weathered	Saturated hydraulic condition Weathered	Modulus of rupture Non-weathered	Modulus of rupture Weathered
	(%)	($g\ cm^{-3}$)		($g\ g^{-1}$)		($cm\ h^{-1}$)		($kg\ cm^{-2}$)	
	0	1.02 c[a]	1.05 c	0.25 b	0.27 b	–	–	0.83 b	1.86 a
Holtville	6.25	1.10 b	1.13 bc	0.24 b	0.30 b	–	–	0.09 b	1.40 ab
clay	12.5	1.14 b	1.15 bc	0.24 b	0.30 b	–	–	0.01 b	1.28 b
	25	1.15 b	1.18 b	0.25 b	0.29 b	–	–	0.01 b	0.72 c
	50	1.26 a	1.27 a	0.33 a	0.36 a	–	–	0.02 b	0.24 b
	0	1.04 b	1.08 a	0.33 a	0.38 a	0.2 b	0.20 a	0.98 a	1.05 a
Willows	6.25	1.04 b	1.14 a	0.33 a	0.42 a	1.6 a	0.01 b	0.24 b	0.84 a
silty clay	12.5	1.07 b	1.14 a	0.34 a	0.42 a	1.7 a	0.01 b	0.12 b	0.77 a
	25	1.09 b	1.14 a	0.35 a	0.42 a	1.9 a	0.01 b	0.08 b	0.56 a
	50	1.13 a	1.14 a	0.37 a	0.41 a	0.1 b	0.02 b	0.05 b	0.12 b
	0	1.37 a	1.37 ab	0.22b	0.21 c	0.3 c	0.34 a	2.22 a	2.35 a
Domino	6.25	1.30 ab	1.43 a	0.22 b	0.23 bc	4.2 ab	0.05 b	0.23 b	0.72 bc
loam	12.5	1.27 b	1.43 a	0.22 b	0.25 abc	5.2 a	0.05 b	0.23 b	0.48 bcd
	25	1.24 b	1.38 ab	0.22 b	0.27 ab	3.4 b	0.03 b	0.06 b	0.28 bcd
	50	1.18 b	1.29 b	0.28 a	0.30 a	1.1 c	0.04 b	0.06 b	0.11 d
Fly ash[b]	100	1.11	1.12	0.44	0.43	0.2	0.2	0.22	0.06

[a] Within each data group, values followed by the same letter indicate they are not significantly different at $P < 0.05$ by Duncan multiple range test.
[b] Not included in the statistical analysis.

5 Summary and Conclusions

1. Solid wastes, such as fly ash and municipal sludges, may be added to soils to alter the physical properties of the soil. The amounts of solid wastes required to accomplish this usually exceeded 25% in the mixtures. Physical behaviors of the waste-amended soils were characterized by enhanced water-holding capacity and hydraulic conductivity, lower bulk density, and less cohesion between soil particles.

2. Major chemical elements of a fly ash produced by a power plant in California were Al, Si, Ca, Mg, Na, and K which accounted for almost 50% of the total solids. Concentrations of trace elements in the fly ash were similar to those typically found in the soils.

3. A survey of 110 municipal wastewater treatments in California indicated that contents of N and P in sewage sludges were related to methods of the wastewater treatment plant and the way sludges were digested. Concentrations of industrial metals in sludges followed a log-normal distribution. Sludges with exceptionally high metal concentrations were from the wastewater which received significant amounts of industrial discharges.

4. Long-term field data indicated that in soils treated with solid waste, a biologically nontoxic element (Na) moved slowly downward in the soil profile, while toxic chemical elements (Cd and Pb) accumulated in the surface soil.

References

Ahrland S, Chatt J, Davis NR (1958) The relative affinities of ligand atoms for acceptor molecules and ions. Q Rev Chem Soc 12:265–76

Amos DF, Wright JD (1972) The effect of fly ash on soil physical characteristics. In: Schnartz MA (ed) Proc 3rd Mineral waste utilization Symp, Mar 5–6, 1972. US Bur Mines IIT Res Inst, Chicago, Ill, pp 75–78

ASTM (American Society for Testing Material) (1970) Procedure for testing soils

Beckett PHT (1980) The statistical distribution of sewage and sludge analysis. Environ Pollut Ser B 1:27–35

Blakeslee PA (1973) Monitoring considerations for municipal waste water effluent and sludge application to land. EPA, USDA, Univ Worksh, Champaign, Urbana, Ill. July, 1973

Bouyoucos GJ (1951) A recalibration of the hydrometer method for making mechanical analysis of soil. Agron J 45:434–438

Bradford GR, Page AL, Straughan IR, Phung HT (1976) A study of the deposition of fly ash on desert soils and vegetation adjacent to a coal-fired generating station. Environ Chem and Cycling Process, Mineral Cycling Symp, Augusta, Ga, USA, April 28-May 1, 1976

Casagrande A (1948) Classification and identification of soils. Trans Am Soc Civil Eng 113:901–930

Chaney R (1987) Toxic element accumulation in soils and crops: protecting soil fertility and agricultural food chain. 2nd Int Worksh Behavior of pollutants in porous media, Bet Dagan, Isr, June 14–19, 1987

Chang AC, Lund LJ, Page AL, Warneke JE (1977) Physical properties of fly ash amended soils. J Environ Qual 6:267–270

Chang AC, Page AL, Warneke JE (1983) Soil conditioning effects of municipal sludge compost. J Environ Eng ASCE 109(3):574–583
Davison RL, Natusch DFS, Wallace JR, Evens CA, Jr. (1974) Trace elements in fly ash: dependence of concentration on particle size. Environ Sci Technol 8:1107–1113
Day PR (1965) Particle fractionation and particle size analysis. In: Black CA (ed in chief) Methods of soil analysis. Agronomy Monogr 9. Am Soc Agron, Madison, Wisc, pp 545–566
Doran JW, Martens DC (1972) Molybdenum availability as influenced by application of fly ash to soil. J Environ Qual 7:69–73
Elseewi AA, Bingham FT, Page AL (1976) Growth and mineral composition of lettuce and Swiss chard grown on fly ash amended soils. Environ Chem and Cycling Process, Mineral Cycling Symp, Augusta, Ga, April 28-May 1, 1976
Elseewi AA, Page AL, Bingham FT (1978) Availability of sulfur in sewage sludge to plants: a comparative study. J Environ Qual 7:213–217
Epstein E (1976) Effect of sewage sludge on some soil physical properties. J Environ Qual 4:139–142
Goecker WL, Moh ZC, Davison PT, Chu TY (1956) Stabilization of fine and coarse grained soils with lime and lime-fly ash admixture. Highway Res Board Bull 129:63–82
Hecht NL, Duvall DS (1975) Characterization and utilization of municipal and utility sludges and ashes, vol 3: Utility coal ash. Environ Protect Technol Ser EPA-670/2-75-033c. US Environ Protect Ag, Washington, DC
Kaakinen JW, Jorden RM, Lawasani MH, West RE (1975) Trace element behavior in coal-fired power plant. Environ Sci Technol 9:862–869
Klein LA, Lang M, Nash N, Kirschner SL (1974) Sources of metals in New York City wastewater. JWPCF 46:2653–2662
Lee RE, Jr., Von Lehmden DJ (1973) Trace metal pollution in the environment. J Air Pollut Control Assoc 23:853–857
Lee RE, Crist HL, Riley AE, MacLeod KE (1975) Concentration and size of trace metal emission from a power plant, a steel plant and a cotton gin. Environ Sci Technol 9:643–647
Logan TJ, Chaney R (1983) Utilization of municipal wastewater on land – metals. In: Page AL, Gleason III TL, Smith JE, Jr., Iskandar IK, Sommers LE (eds) Proc 1983 Worksh Utilization of municipal wastewater and sludges on land. Univ Cal, Riverside, pp 235–326
Natusch DFS, Bauer CF, Matusiewicz H, Evans CA, Baker J, Loh A, Linton RW, Hopke PK (1975) Proc Int Conf Heavy metals in the environment, Toronto, Ontario, Can, vol 2, pt 2, pp 553–576
Nieboer E, Richardson DH (1980) The replacement of nondescript term heavy metals by a biologically and chemically significant classification of metal ions. Environ Pollut Ser B 1:3–26
Ochiai E (1977) Bioinorganic chemistry: an introduction. Allyn & Bacon, Boston, MA
Phung HT, Lund LJ, Page AL (1976) Potential use of fly ash as a liming material. Environ Chem and Cycling Process, Mineral Cycling Symp, Augusta, Ga, April 28–May 1, 1976
Plank CO, Martens DC (1974) Boron availability as influenced by application of fly ash to soil. Soil Sci Soc Am Proc 38:974–977
Reeves RC (1969) Modulus of rupture. In: Black CA (ed) Methods of soil analysis, vol 1. Am Soc Agron, Madison, Wisc, pp 466–471
Richards SJ, Weeks LV, Warneke JE (1965) A soil compactor and procedure used for preparing cores for measuring physical properties. Soil Sci Soc Am Proc 29:637–639
Sommers LE (1977) Chemical composition of sewage sludges and analysis of their potential use as fertilizers. J Environ Qual 6:225–232
Swanson VE, Medlin JM, Hatch JR, Coleman SL, Wood GH Jr, Woodruff SD, Hildebrand RT (1976) Collection, chemical analysis and evaluation of coal samples in 1975. US Dep Interior, Geol Surv, Open File Rep 76:468, 503 pp
Tabatabai MA, Frankenberger WT Jr (1979) Chemical composition of sewage sluges in Iowa. Iowa State Univ Res Bull 586:933–944
Watt JD, Thorne DT (1965) Composition and pozzolanic properties of pulverized fuel ashes: I. Composition of fly ash from some British power stations and properties of their component particles. J Appl Chem 15:585–594

Toxic Element Accumulation in Soils and Crops: Protecting Soil Fertility and Agricultural Food-Chains

R.L. Chaney[1]

1 Introduction

Assessment of the likelihood of risks to crop productivity, humans, livestock, and wildlife from potentially toxic elements in sewage sludge applied to land (and other metal sources such as fertilizers, manures, and stack emissions) requires a knowledge of the potential for transfer of each element from the sludge or sludge-soil mixture to crops and to animals (including humans) which ingest sludge, sludge-soil mixtures, or crops grown on the sludge-amended soil, including the bioavailability of the element as ingested. Great progress has been made in characterizing the potential for excessive food-chain transfer of potentially toxic elements, and the risk of reducing soil fertility due to metal toxicity.

Beginning in 1980, two new concepts in risk assessment for sludge-applied trace elements were introduced (Chaney 1980). One was the "Soil-Plant Barrier", and the other was the great importance of interactions among elements which strongly affect element bioavailability or toxicity. These interactions increase or decrease the potential for element toxicity. Traditional toxicology considers only one element at a time, usually adding a soluble salt of that element to a minimal diet to maximize bioavailability and interactions. However, we are interested in the potential risk from applying a complex organic waste to soils, and element transfer to crops and to animals fed practical diets. We must consider the possible food-chain transfer and toxicity of the sludge-applied element, and this case is usually quite different from the traditional toxicology case. I have called this approach "Environmental Toxicology", because it deals with real cases of metal enrichment of soils in practical agricultural ecosystems. We still consider the "worst case" conditions, and protect the "most exposed, most sensitive individual", but constrain these to realistic agricultural and environmental conditions. As will be noted below, the concentration of an element in a sludge is usually very important in estimating whether it can cause risk, and industrially polluted municipal sludges must be carefully evaluated and sludge composition regularly monitored to avoid risk. Table 1 shows the range of concentrations found in sludges for important elements; typical median levels and maximum domestic levels are also noted. Nearly all problems from sludge metals require the element to be present at levels far greater than median concentrations.

[1]USDA-Agricultural Research Service, Soil-Microbial Systems Laboratory, B-318, BARC, Beltsville, MD 20705, USA

Table 1. Range in concentrations of selected trace elements in dry digested sewage sludges[a] (updated from Chaney 1983)

Element	Reported range		Typical median sludge	Typical soil	"Maximum domestic sludge
	Minimum	Maximum			
As, mg kg^{-1}	1.1	230	10	-	-
Cd, mg kg^{-1}	1	3,410	10	0.1	25
Cd/Zn, %	0.1	110	0.8	-	1.5
Co, mg kg^{-1}	11.3	2490	30	-	200
Cu, mg kg^{-1}	84	17,000	800	15	1000
Cr, mg kg^{-1}	10	99,000	500	25	1000
F, mg kg^{-1}	80	33,500	260	200	1000
Fe, %	0.1	15.4	1.7	2.0	4.0
Hg, mg kg^{-1}	0.6	56	6	-	10
Mn, mg kg^{-1}	32	9,870	260	500	-
Mo, mg kg^{-1}	0.1	3,700	20	-	35
Ni, mg kg^{-1}	2	5,300	80	25	200
Pb[b] mg kg^{-1}	13	26,000	250	15	500
Sn, mg kg^{-1}	2.6	329	14	-	-
Se, mg kg^{-1}	1.7	17.2	5	-	-
Zn, mg kg^{-1}	101	49,000	1700	50	2500

[a] Composting using wood chips as a bulking agent generally produces composted sludge 50% as high in trace elements as a digested sludge from the same treatment plant.
[b] Sludge Pb concentration has dropped significantly during the 1980's in the USA due to reduction of Pb in gasoline.

In more recent publications, we provided greater detail for risk assessment of the potentially toxic elements in sewage sludge (Chaney 1980, 1983, 1985; Logan and Chaney 1983; Hansen and Chaney 1984; Chaney et al. 1987). Although a brief summary of these concepts is provided here, the reader is referred to other papers for more details and further references.

In recent years, other concepts have been developed which appear to be another important improvement in assessing the long term risk of sludge-applied trace elements. In particular, the plateau (nonlinear) model for plant response to sludge-applied metals appears to have a general application in risk assessment for waste-borne heavy metals such as Cd, Zn, and Cu. The most recent advance to be summarized is a new method for quantitative estimation of the potential for Cd transfer from soil to humans under worst case conditions.

2 Transfer of Sludge-Applied Trace Elements to Animals by Direct Ingestion of Sludge or Sludge-Amended Soil

As research on risk from sludge utilization progressed, it became clear that sludge-applied elements are most readily transferred to humans and livestock by direct ingestion of sludge adhering to vegetation or lying on the soil surface. Sludge is often utilized on pastures because the land is accessible intermittently during the growing season. Spray-applied sludge can adhere to forage crops and can allow direct ingestion of the sludge particulates. Alternatively, surface-applied sludge materials become part of the organic rich "thatch layer" on the soil surface. Grazing livestock inadvertenly ingest surface soil (Fries 1982; Fries et al. 1982a, b). Sludge products (e.g., compost) are surface-applied on lawns, gardens, and perennial plantings. Children ingest dust and soil as part of their normal play activities. These processes can allow much more direct exposure to organic and inorganic constituents in sludge than can occur via plant uptake.

Several studies have evaluated crop and sludge factors in adherence of spray-applied sludge to forage crops. Increased sludge application rate (within practical rates) increased sludge on the forage (Chaney and Lloyd 1979). If sludge dried on the crop, it largely remained there regardless of rainfall. Jones et al. (1979) found that sludge could be washed off immediately after application, but not after it dried. They further showed that sludges with higher solids content caused more sludge adherence, perhaps because of viscosity. Chaney et al. (1987) summarized other research on crop differences in adherence of spray-applied sludge. These differences between crop species are small compared to the factors described above.

Although forages can reach 15–30% sludge (dry matter basis) immediately after spray application, the US-Environmental Protection Agency regulations (EPA 1979a) now require a 30-day waiting period before grazing; users are advised to spray-apply sludge only to well-grazed or clipped forages. This requirement reduces the initial adherence and growth rapidly dilutes the adhering sludge. In research under practical grazing management conditions, Decker et al. (1980) and Bertrand et al. (1981) found that sludge comprised only 2–3% of the dry diet (based on forage and feces analyses; Table 2). Injection of sludge in the plow layer of soil prevents sludge adherence to crops, and greatly reduces the potential ingestion of sludge on the soil. Subsurface injection of fluid sludges is now considered the best management practice, although median metal concentration sludges may not represent risk when spray-applied under good management.

Ingestion of sludge from the soil surface was estimated for a compost amendment which did not adhere to the crop (Decker et al. 1980). Cattle consumed about 1–3% compost when the forage had no detectable compost adhering (Table 2). Lower compost ingestion occurred in 1978 when compost was applied only once during the growing season.

Grazing livestock have been found to ingest 1–40% soil (Chaney 1983). High soil consumption generally results from poor management, overgrazing, and/or trampling forage into moist soil. Others have evaluated soil consumption by well-managed dairy cattle, sheep, and swine (Fries et al. 1982a, b). Up to 8% soil was ingested from pasture, and less from bare soil. Hogue et al. (1984) reported metal residues in sheep tissues after the sheep grazed 152 days on a grass-legume pasture established on soil in which 224 Mg ha^{-1} metal-rich sludge had been

Table 2. Adherence of spray-applied liquid sewage sludge to tall fescue (Decker et al. 1980) or "Pensacola" bahaigrass (Bertrand et al. 1981) and sludge content of feces of cattle which rotationally grazed these pastures (see Chaney et al. 1987 for details)

Study and treatment	Sludge solids (%)	Application rate[a] (cm) (No.)	Sludge in/on forage (%)	Sludge in feces (%)
Decker et al. (1980)				
1976: 21-day sludge	4.4	0.51 × 20	5.39	7.1
1976: 1-day sludge	4.8	0.51 × 20	22.3	18.6
1977: 21-day sludge	2.9	0.51 × 20	2.18	7.7
1977: Compost			(0.74)[b]	6.5
1978: 21-day sludge	3.7	0.51 × 24	2.91	6.1
1978: Compost			(0.50)[b]	2.0
Bertrand et al. (1981)				
1979: 7-26 day sludge	2.1	0.84 × 9	2.17	4.6
1979: 7-13 day sludge	2.1	0.84 × 18	5.17	5.8

[a] Application rates in cm of liquid sludge and number of times applied during season.
[b] Composted sludge did not significantly adhere to forages.

incorporated. Although the forage was increased in Cd, sheep kidney-Cd was increased insignificantly and other element residues were not influenced by sludge incorporation. Similar results were obtained when sheep grazed pastures which received surface-applied, high Cu, swine manure during the previous grazing season. Neither liver nor fecal Cu were consistently affected by previous manure application (Poole et al. 1983).

2.1 Bioavailability of Elements in Ingested Sludge

Ingestion of sludge-borne trace elements does not necessarily cause the health effects which would be expected from traditional toxicological studies with added metal salts. Sludge-feeding studies have been conducted to evaluate element deposition in tissues of cattle, sheep, and swine. Low metal sludges have not increased Cd, Zn, Pb, or Cu in animal tissues in several studies (Decker et al. 1980; Baxter et al. 1982; Sanson et al. 1984; Smith et al. 1985). However, high Cd sludges have increased Cd in liver and kidney (Kienholz et al. 1979; Bertrand et al. 1981; Hansen et al. 1981; Johnson et al. 1981; Baxter et al. 1982; see review in Hansen and Chaney 1984). The most consistent potential problem resulting from sludge ingestion is reduced Cu concentration in the liver. Sludge Zn, Cd, Fe, sulfide, and possibly Mo could interfere with Cu absorption. Ingestion of sludge containing 11% Fe induced Cu-deficiency in cattle in the only sludge-feeding or sludge-grazing study where animal performance or health was negatively affected (Decker et al. 1980). When *ingested* sludges are rich in Cd, Hg, F, or Pb, deposition occurs in bone or liver, but little change has been found in livestock tissues used as food (Hansen and Chaney 1984). The Soil-Plant Barrier concept indicates that plant uptake of Hg, F, or Pb cannot cause toxicity in livestock; rather, direct ingestion of sludge rich in these elements is required for toxicity to, or excessive tissue burdens in livestock.

Crops grown on sludge-fertilized soils can transfer trace elements to feeds and foods. However, the extent of increases of trace elements in crop tissues, and the bioavailability of these to animals, varies with sludge properties. Crops fertilized with low metal sludges had little effect on kidney Cd in several studies; however, higher Cd sludges increased Cd in crops, which had no effect other than increasing Cd in kidney and liver (Chaney et al. 1987; Hinesly et al. 1985; see below). Further, sludge fertilized crops have contained five times higher Cd than control crops, yet did not increase kidney Cd above the control (Chaney et al. 1978a, b; Evans et al. 1979; Boyd et al. 1982; Bray et al. 1985).

3 "Soil-Plant Barrier" Protects the Food-Chain Against Most Toxic Elements

When present in soils, some elements are easily absorbed by plants and are translocated to food-chain plant tissues (e.g., Zn, Cd, Mn, Mo, Se, B), while others are not. These other elements are strongly bound to soil or retained in plant roots, and are not translocated to plant foliage in amounts injurious to livestock, even when soils are greatly enriched (e.g., Fe, Pb, Hg, Al, Ti, Cr^{3+}, Ag, Au, Sn, Si, Zr). Although an element may be relatively easily absorbed and translocated to plant foliage, phytotoxicity may limit plant levels of these elements to concentrations safe for animals (e.g., Zn, Cu, Ni, Mn, As, B).

During the last 40 years, these concepts were developed by many researchers (see Chaney 1983). Chaney (1980) introduced the term "Soil-Plant Barrier" to describe these concepts when considering waste-soil-plant-animal relationships of potentially toxic microelements. A soil-plant barrier protects the food-chain from toxicity of a microelement when one or more of the following processes limit the maximum levels of that element in edible plant tissues to levels safe for animals: (1) insolubility of the element in soil prevents uptake; (2) immobility of an element in fibrous roots prevents translocation to edible plant tissues; or (3) phytotoxicity of the element occurs at concentrations of the element in edible plant tissues below that injurious to animals. Unfortunately, the soil-plant barrier does not protect animals from toxicities of all elements. The exceptions important in assessing risk from land application of municipal sludge are Cd, Se, and Mo; a few more elements may have to be considered for land application of industrial wastes (Be, Co).

Ingestion of amended soil or sludge can circumvent the soil-plant barrier. Many elements are so insoluble or nontoxic that animal health is not influenced even if ingested soil or waste contains the element (e.g., Cr^{3+}, Zr, Ti, Al, Sn, Si). However, direct ingestion of sludges with concentrations well in excess of median levels in some elements such as Cu, F, Zn, Pb, Fe^{2+}, As, Co and Hg causes risk to livestock. These risks would have been insignificant if the sludge were mixed with the surface soil (0–15 cm).

4 Interactions Among Dietary and Sludge Constituents Influence Microelement Impact on the Food-Chain

Evaluation of the potential impact of microelements on animals via their consumption of sludge, sludge-amended soil, or crops grown on sludge-amended soil, is very complex. Animal species differ in tolerance of microelements. Tolerance to microelements is also influenced by age; younger animals are generally more sensitive than older. Crop species absorb unequal amounts of microelements. Total and relative microelement uptake is affected by crop species and cultivar, soil pH, organic matter, soil temperature, and other factors. Wastes differ in levels of elements and ratios among elements. Individual, potentially toxic elements interact with other elements in the diet, often reciprocally. These interactions are often the basis for physiological toxicity; hence, interactions or microelement imbalance are of great importance in assessing risk.

When one element is so increased in animal diets that the ratio of it to other elements or dietary constituents is great enough to induce a deficiency of another, then animal weight gain declines and/or a health effect is observed. For example, Zn toxicity to cattle is normally not observed until dietary Zn exceeds 500 mg kg^{-1} dry diet. However, if the diet is marginal or deficient in Cu, as little as 300 mg Zn kg^{-1} can injure sheep (Campbell and Mills 1979). Domestic sewage sludge contains a mixture of potentially toxic elements. Consumption of sludge or sludge-amended soil is a very different case for risk assessment than standard toxicological studies where a soluble salt of one element is added at rates causing health effects (and often to purified rather than practical diets). With sludge ingestion, increased levels of dietary Zn are balanced by increased levels of Cu and Fe. Thus, the Cu deficiency required for the "most sensitive" case for Zn toxicity to livestock does not exist when sludge supplies Zn to soil. Also, in contrast to Cu salts mixed with diets, Cu in sludge fed to livestock has low bioavailability. Generally, ingested sludge lowers liver Cu stores rather than causing Cu toxicity, although equal levels of soluble Cu salts would poison the animals (Decker et al. 1980; Bertrand et al. 1981; Baxter et al. 1982). Recently, research on potential toxicity from ingestion of high Cu swine manure has led to the same conclusion, i.e., interactions can reverse toxicity predicted from "toxicology" studies (Bremner 1981; Poole et al. 1983). A number of elements are considered in detail regarding sludge-soil-plant-animal interactions influencing the food-chain in other publications (Chaney 1980, 1983, 1985; Logan and Chaney 1983; Hansen and Chaney 1984), and are summarized in Table 3. Cadmium is discussed below.

Table 3. Maximum tolerable levels of dietary minerals for domestic livestock in comparison with levels in forages

Element	"Soil-Plant Barrier"	Level in plant foliage[a] Normal	Phytotoxic	Maximum levels chronically tolerated[b] Cattle	Sheep	Swine	Chicken
		(mg kg^{-1} dry foliage)		(mg kg^{-1} dry diet)			
As_{inorg}	Yes	0.01–1	3–10	50	50	50	50
B	Yes	7–75	75	150	(150)	(150)	(150)
Cd[c]	Fails	0.1–1	5–700	0.5	0.5	0.5	0.5
Cr^{3+}	Yes	0.1–1	20	(3000)	(3000)	(3000)	3000
Co	Fails	0.01–0.3	25–100	10	10	10	10
Cu	Yes	3–20	25–40	100	25	250	300
F	Yes?	1–5	–	40	60	150	200
Fe^{2+}	Yes	30–300	–	1000	500	3000	1000
Mn	?	15–150	400–2000	1000	1000	400	2000
Mo	Fails	0.1–3.0	100	10	10	20	100
Ni	Yes	0.1–5	50–100	50	(50)	(100)	(300)
Pb[c]	Yes	2–5	–	30	30	30	30
Se	Fails	0.1–2	100	(2)	(2)	2	2
V	Yes?	0.1–1	10	50	50	(10)	10
Zn	Yes	15–150	500–1500	500	300	1000	1000

[a] Based on literature summarized in Chaney (1983).
[b] Based on NRC (1980). Continuous long-term feeding of minerals at the maximum tolerable levels may cause adverse effects. Levels in parentheses were estimated (by NRC) by extrapolating between animal species.
[c] Maximum levels tolerated were based on Cd or Pb in liver, kidney, and bone in foods for humans rather than simple tolerance by the animals.

An important reference for tolerance of microelements by animals has been published by the US National Research Council (NRC 1980). The NRC committee considered increased levels of only the element being evaluated, although they discussed interactions. Their tolerance levels are shown in Table 3. Unfortunately, these levels often are not valid for sludge-fertilized crops or for ingestion of sludge or soil because of the noted interactions.

5 Assessing Potential Risk from Sludge-Applied Cadmium

Much sludge risk assessment has focused on Cd, because Cd can be mobile in food-chains, and because soil Cd had caused human health effects in Japan. Many aspects of Cd food-chain transfer, and Cd toxicity to humans are complicated, and it now appears that sludge Cd poses much lower risk than thought in the early 1970's.

5.1 Cd Bioavailability

One source of error identified is represented by methods used to assess Cd bioavailability using humans and laboratory animals. Problems have been identified with the experimental methods used in this research. Early studies (Rahola et al. 1973; Yamagata et al. 1975) measured retention after only a few days or weeks. McLellan et al. (1978) found that part of the diet Cd was absorbed by intestinal mucosal cells. These were subsequently sloughed into the intestine and the diet Cd repeatedly recycled in intestinal cells for a prolonged time. Fox et al. (1984) showed that this aspect of Fe deficiency allows an increase in true Cd absorption and movement to kidney long after other parts of the test diet residue are excreted.

Flanagan et al. (1978) found that Fe sufficiency very strongly reduced Cd retention. The Fe sufficiency allows ferritin in the intestinal mucosal cells to adsorb Cd absorbed into them (Huebers et al. 1987). Shaikh and Smith (1980) were able to study subjects for up to 800 days after the test dose (using ^{109}Cd), and resolved whole body Cd into three pools, including the slowly excreted intestinal Cd turnover pool. The biological half-life of the slowest pool was 18 years to infinity rather than 100 days as previously reported for the shorter-term human studies. Again, Fe deficiency increased retention of dietary Cd, but the mean retention was appreciably lower than in earlier studies (2.8%, Shaikh and Smith 1980 vs 5.3%, Flanagan et al. 1978). These tend to support the findings of Newton et al. (1984), Sharma et al. (1983), and McKenzie-Parnell and Eynon (1987) that only low amounts of Cd are retained by humans ingesting Western-type diets.

The cited animal studies also show that dietary Zn and Ca are also important in Cd absorption. Increased Zn apparently induces biosynthesis of metallothionein in the intestinal mucosal cells; this protein then binds both Cd and Zn. Zn may also inhibit Cd absorption into these cells. Calcium deficiency causes the intestine to increase its ability to absorb Ca, which also increases Cd absorption. Like Zn, Ca may also compete in the absorption process directly.

Because of these interactions, and because of the unique characteristics of rice which carried soil Cd to the Japanese farm families, humans suffered the Fanconi syndrome, a proximal renal tubular proteinuria caused by Cd, and some women suffered "itai-itai" disease, a painful osteomalacia induced by chronic Cd poisoning under conditions of low dietary Ca.

Rice was the most important food source of Cd consumed by these farm families. The rice was grown on soils rich in both Cd and Zn due to contamination by Zn-ore wastes. In contrast to other food crops, rice is grown in flooded soils. If sufficient S is present, and enough energy available, Cd and Zn uptake by rice is kept very low because insoluble metal sulfides are formed in the soil. However, some metal uptake occurs because the roots obtain oxygen by air channels within the plants, and the farmers drain the fields early to obtain higher yields and to facilitate harvest. It was found that rice metabolized Cd and Zn differently in these anaerobic rice paddy soils than patterns common to other food crop species (e.g., wheat, lettuce) in aerobic soils. Although both soil Cd and soil Zn were greatly increased (up to 10 mg Cd kg^{-1} and 1200 mg Zn kg^{-1}), rice grain Cd increased up to 100-fold while grain Zn was unchanged (Fukushima et al. 1973). Further, during preparation of polished rice, much of the Ca, Zn, and Fe in brown rice are removed during milling (Table 4), while a much lower fraction of the Cd is removed (Pedersen and Eggum 1983; Chino and Baba 1981). Lastly, rice Fe has very low bioavailability (Layrisse and Martinez-Torres 1971; Hallberg et al. 1977). All these factors [crop metal uptake characteristics, food processing, and effect of nutrient (Fe, Zn, and Ca) status on human Cd retention] favored Cd retention by the farmers. Bioavailability of Cd was thus high, and human disease resulted.

Table 4. Effect of milling brown rice on minerals remaining in rice used as food (dry weight basis) (based on Pedersen and Eggum 1983)

	Type of rice				
Mineral	Rough	Brown	Under-milled	Milled	Over-milled
Flour yield, %	100	82	77	72	64
Ca, mg g^{-1}	0.31	0.12	0.10	0.08	0.07
Phytate-P, mg g^{-1}	2.1	2.2	0.70	0.20	0.10
Zn, $\mu g\ g^{-1}$	24	33	20	18	16
Fe, $\mu g\ g^{-1}$	38	8.8	7.0	4.1	2.4

In another Cd exposure case, individuals who consumed large amounts of Cd-rich oysters had higher dietary levels of Zn, Fe, and Ca, levels which were more like those in the normal US diet. The oyster Cd had very low apparent bioavailability based on Cd in blood and urine compared to the effect of smoking on Cd in blood and urine (Sharma et al. 1983; McKenzie-Parnell and Eynon 1987; McKenzie-Parnell et al. 1988). Nordberg et al. (1986) found that most of the Cd in the oysters was bound to high molecular weight proteins, and not as a metallothionein as is found when acute doses are fed to oysters; further, the oysters contained 5.3 mg Cd and 100 mg Zn kg^{-1} wet weight. McKenzie-Parnell and Eynon (1987) have recently confirmed, by kidney analysis in deceased oyster consumers, that Cd absorption has been quite low in this population. These findings should reduce the predictions of human retention of Cd from Western-type diets.

5.2 Crop Cd Relationship with Soil Cd

Transfer of sludge-applied Cd from soil to crops is predominantly a function of: (1) the Cd concentration in the applied sludge; (2) soil pH; (3) characteristics of the applied sludge (especially Fe concentration) and cumulative sludge application rate; and (4) the crop species and cultivar grown. Each of these factors can change crop Cd uptake by more than two fold.

Earlier attempts to estimate food-chain transfer have used plant uptake slopes obtained by linear regression of the Cd concentration in edible crop tissue on the amount of the Cd applied [(mg constituent kg^{-1} dry crop) per (kg constituent applied ha^{-1})]. However, it has recently been shown that, in the long-term, plant uptake of sludge-applied Cd and Zn is a curvilinear (plateauing) response to the cumulative application rate of an applied constituent in a sludge rather than the simple linear response presumed by many (Chaney et al. 1982; Corey et al. 1987). Further, the height of the plateau above the Cd level in the control crop increases as the concentration of the Cd in sludge increases, and decreases as the amount of Fe added during sludge processing increases. This information indicates an improved approach to regulating sludge application wherein the sludge "quality" is important. Figure 1 shows the concentration of Cd in Romaine lettuce grown on a Christiana sandy loam amended once with several levels of two sludges; separate plots were adjusted to strongly acidic and near-neutral soil pH. Lettuce was grown for 8 years. Orthogonal contrasts showed that the application-rate-squared term was significant, indicating that the simple

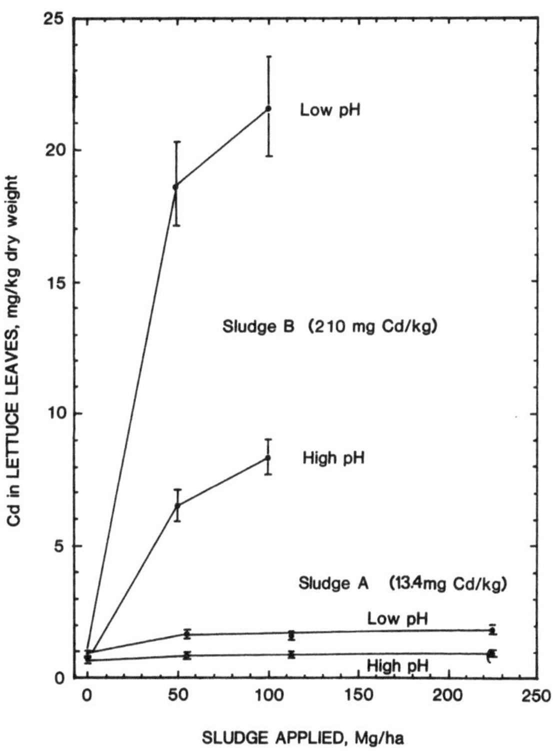

Fig. 1. Effect of sludge application rate on Cd in lettuce leaves. *Sludge A* (13.4 mg Cd kg^{-1} and 8.3% Fe) was applied in 1976 and lettuce grown from 1976 to 1983. *Sludge B* (210 mg Cd kg^{-1} and 2.5% Fe) was applied in 1978 and lettuce grown in 1978 to 1983. Results shown are geometric means over years ± one SE

linear response model was not an adequate model of the soil:crop Cd transfer. Other research has supported this curvilinear model (Corey et al. 1987). The current explanation is that sludge applies both Cd and specific Cd sorption capacity to the soil; after soil-specific Cd sorption sites are filled with sludge-applied Cd, the sludge sorption capacity controls Cd availability to crops. Thus a plateau model is expected, and is found by use of a nonlinear regression on the above data. The increase in lettuce Cd above the control treatment will be used in a later section to estimate dietary Cd increase for gardeners.

5.3 Long-Term Availability of Sludge-Applied Cd to Crops

Although there has been much discussion of reversion (reduction over time) of the availability of soil metals, there has been little demonstration or explanation of this result. The workshop report by Chang et al. (1987) concludes that two types of reversion are observed: (1) the stabilization of biodegradable organic matter causes lower levels of soluble organic ligands in the soils solution, which reduces metal convection and diffusion and thereby reduces plant uptake 1 to 2 years after high rates of sludge are applied (see also Logan and Chaney 1983); and (2) slow long-term reaction of metals with the soil. This latter phenomenon is partially explained by research of Bruemmer et al. (1986) who found that Zn, Cd, etc. could diffuse into goethite (an Fe oxide persistent in soils) with time. They also showed the great specific sorption capacity of Fe, Mn, and Al oxides for Cd and Zn. They concluded: "The slow diffusion of trace or toxic heavy metals into soil minerals under natural conditions may in part explain why heavy metals added to soils as soluble salts or deposited by pollution processes become immobilized and less available with time."

6 Estimating Cd Transfer from Sludge-Amended Soils to Human Diets

Estimation of food-chain transfer is critical to valid estimation of the potential for risk. At present, these estimates are best made by considering (1) the relative increased uptake of Cd by various crops from sludge-amended soil under soil conditions where substantially increased uptake occurs; (2) the rate of ingestion of different crops by the US population (e.g., Pennington 1983); (3) the demonstrated bioavailability of the increased amounts of Cd in sludge-grown crops or ingested sludge; and (4) the new appropriate transfer coefficient from sludge-amended soil to edible crop tissues [i.e., the increase in crop trace element residue (above that in the same crop grown on background soils) at the plateau reached on sludge-amended soil].

Ryan et al. (1982) developed an approach for estimating food-chain transfer of sludge-applied Cd. While this model has to be modified to account for the curvilinear response to sludge-applied Cd, many other parts of this approach remain valid. In particular, the aggregate increased food-chain exposure to a sludge-borne constituent for the exposed population can be expressed in terms of a constant times the element transfer coefficient (height of the plateau above that for untreated soil) for an indicator crop such as lettuce. Individuals who grow a significant portion of the garden vegetables they ingest on acidic sludge-amended soils, are generally believed to be those most likely to have increased dietary Cd due to land application of sludge. For chronic, lifetime (50-year) Cd exposure, the estimation also relies on US food intake estimates (g fresh weight day^{-1}) from Pennington (1983), and solids content of foods from USDA data bases (Adams 1975). Because response curves or plateaus have not yet been evaluated for every crop consumed by US citizens, uptake by various food crops must be represented by the FDA food groups. For Cd, based on many data sources, the response of minor food crops remains well represented by the FDA food groups.

On the other hand, the use of a "Cd-accumulator" crop to represent increased Cd uptake by all crops in a food group has been criticized. For example, lettuce and cabbage were both listed in the leafy vegetable food group, but have at least ten fold different response slopes. Root vegetables and garden fruits also include high and low element accumulating crops. These wide crop differences have caused an excessive estimated increase in food-chain Cd transfer (providing a hidden safety factor). Thus, food group aggregate Cd transfer-response slopes have to be adjusted for the proportion of low and high Cd-accumulating crops in each food group. Each food group can still be represented by a constant times the response of lettuce. This approach appears to be the most appropriate available based on current research findings.

Many studies have shown significantly higher plant uptake of Cd, Zn, and Ni at lower soil pH compared to neutral soil pH (see CAST 1980; Logan and Chaney 1983). Exceptions were reported in which corn silage was unchanged or slightly higher in Cd on limed than unlimed sludge-amended soil (e.g., Pepper et al. 1983). One possible reason for this exception is that corn differs from other crops in its mechanism of obtaining Fe from soil. Recent research on the mechanism which plants use to obtain Fe from soil has shown that the Poaceae (corn, wheat, barley, rice, oat, sorghum, etc.) excrete an organic chelating agent which facilitates Fe diffusion to the root and Fe uptake (Sugiura and Nomoto 1984; Romheld and Marschner 1986), while plants in other families do not excrete chelators. These compounds also chelate Cd, Cu, Zn, and other microelements in the presence of Fe (unlike bacterial siderophores, which have very high selectivity for ferric iron). Nonspecific chelating agents added to soils are also known to facilitate diffusion and plant uptake of Zn, Cd, and other elements. Thus, Cd uptake by Poaceae can have a weaker response to soil pH change than other species, depending on the availability of soil Fe.

In any assessment of food-chain transfer of potentially toxic trace elements to humans, the *risk to individuals* ("most exposed", "most sensitive individuals") must be considered. Important problems remain in estimating the variance in potential risk due to unusual individual dietary selection patterns, and to individual variation in retention of potentially toxic constituents in foods. For example, it would be useful to know the statistical distribution of increased Cd intake among individuals consuming self-selected diets using crops grown on sludge-amended acidic gardens. If these data were available, one could design Cd limits to protect individuals with the 95th or 99th percentile of increased exposure. Several studies have noted that the relevant information needed on variation in food intake for the evaluation of *chronic* risk from food Cd is lifetime (50-year) variation in intake of foods, rather than the 1-day variation now available (Dean and Suess 1985). Although the mean 1-day ingestion of foods for a population is estimated well by a large survey of individual 1-day intakes, the variance in the long-term average daily intake by an individual is greatly overestimated by the variance of 1-day intakes for a population. Beaton et al. (1983) and Sempos et al. (1985) found that intraindividual variance was greater than the interindividual variance for multiple 1-day observations. Much smaller variances among individuals are associated with long-term dietary Cd intakes, especially if one considers major dietary subgroups (e.g., vegetarians) separately. Thus, present approaches for estimating the fraction of the population ingesting more Cd than some limit (e.g., 95th percentile, or fraction $> 71\ \mu g$ Cd day^{-1}), based on variances in population 1-day food intakes, would greatly overestimate the lifetime Cd intake by individuals.

A practical approach to estimate the relation between sludge-applied Cd, for example, and increased dietary Cd exposure, is to integrate all crop responses in terms of Cd uptake by a responsive reference crop such as lettuce. The relative increased Cd uptake among crops grown on soils with increased plant available Cd due to sludge application has been relatively consistent among a number of field research studies. However, few individual experiments have included a large number of crop species in a responsive sludge treatment to provide the relative increases in crop levels of potentially toxic constituents needed for dietary exposure assessments.

A recent workshop report (Chaney et al. 1987) summarizes an improved method to estimate food-chain transfer of Cd. Data for relative crop uptake of Cd in many published studies were evaluated. Table 5 shows the summarized food group relative increased Cd uptake.

Although it has been clearly shown that crops differ in uptake of trace elements from the same soil, the biochemical/physiological basis for crop differences has not been identified. Basic research by Jarvis et al. (1976) indicated that crops differed in Cd uptake by roots, and also differed in the fraction of root Cd translocated to shoots. Recently, Grill et al. (1985) found that many plant species made a family of cysteine-rich peptides (related to glutathione) which strongly chelate Cd, Zn, Cu, Pb, and Zn. These "phytochelatins" are synthesized in fibrous roots, chelate absorbed metals, and thereby protect root metabolism and reduce trace element translocation to edible plant tissues (Grill et al. 1985; Robinson and Jackson 1986).

Table 5. Food group aggregation of food intake results, and Cd uptake by food groups relative to that of lettuce (= 1.00), assuming strongly acidic soil. Data (Pennington 1983) for food consumption by six age x sex groups (ages 14–65) were averaged; wet weight conversion to dry weight conducted on individual food basis using data from Adams (1975). Relative uptake of Cd from Chaney et al. (1987)

Food group[a]	Food intakes Wet day	Dry day	Dry weight	Relative increased Cd uptake	Relative increased Cd intake
	(g day^{-1})		(%)		
Leafy vegetables-high	22.376	1.121	5.0	1.00	1.121
Leafy vegetables-low	12.966	1.284	9.9	0.13	0.167
Potatoes	66.438	24.063	36.2	0.020	0.481
Root vegetables-high	3.40	0.605	17.8	0.20	0.121
Root vegetables-low	10.60	1.437	13.6	0.052	0.075
Legume vegetables	42.39	12.640	29.8	0.010	0.126
Garden fruits-high	35.537	3.319	9.3	0.020	0.066
Garden fruits-low	31.949	4.671	14.6	0.010	0.047
					2.20

[a] Leafy vegetables-high includes lettuce and spinach; leafy vegetables-low includes cabbage, kale, broccoli, etc.; root vegetables-high includes carrots; root vegetables-low includes radish, turnip, beet, onion, and leek; garden fruits-high includes tomato products and pepper; garden fruits-low includes cucurbits, sweet corn, and strawberries.

7 Estimating Maximum Allowable Soil Cd Loading Based on Predicted Increase in Dietary Cd

Several methods have been used in different nations and at different times to estimate the maximum cumulative soil Cd, or sludge Cd application which protects the health of individuals (Dean and Suess 1985). This is a very complex issue, as has been noted by Ryan et al. (1982) and Logan and Chaney (1983). The analysis given in Ryan et al. (1982) was considered when the US EPA proposed the existing regulations on land application of sludge in 1979 (Environmental Protection Agency 1979a). A background document (Environmental Protection Agency 1979b), reporting the scientific basis for the regulations, was released at the time the interim final regulations were published.

Based on FDA dietary Cd intake estimates (39 μg Cd day^{-1} for teenaged males), and WHO/FAO recommendations for maximum tolerable weekly Cd intake (52–71 μg Cd day^{-1} for 60 kg humans) [or the Ryan et al. (1982) suggested limit of 150 μg day^{-1} minus 75 μg Cd to allow for smoking three packs of cigarettes per day], EPA (1979a) concluded that sludge could safely add no more than 30 μg Cd day^{-1} to an individual's diet. The high-risk or high-exposure individual was to be protected by the regulation: "That high-risk situation is one where an individual receives 50% of his vegetable diet from sludge-amended soils for a period of 40 to 50 years." The US-EPA recognized the strong effect of soil pH on Cd uptake by crops. For soils with low background pH, it was considered likely that soil pH would fall (from the pH 6.5 required during the permitted period of sludge application) to background soil pH.

Thus, if lettuce were increased by 1 mg Cd kg^{-1} dry weight, garden foods were estimated to be increased by 7.90μg Cd day^{-1} for 100%, or 3.95 μg day^{-1} for 50% of garden foods grown in an acidic, sludge-amended garden for 40–50 years (EPA 1979a). EPA judged that strongly acidic soils (pH 5.4 to 6.2) would not cause greater than 30μg Cd increase day^{-1}, although very acid soils (4.9) caused larger increases. By 1981, several errors in this approach were identified, and an attempt was made to correct these errors. By using adult food intakes rather than that of teenaged males, the estimated increase was only 3.79 μg Cd day^{-1} (for 100% of garden vegetables grown on acidic, sludge-amended soil) when lettuce is increased by 1 μg Cd g^{-1} dry weight. If one divides 35 μg Cd day^{-1} by 3.79 μg increased dietary Cd (per 1 μg Cd increase/g dry lettuce), one finds leafy vegetables could safely reach 9.23 μg Cd increase/g for 100% of diet; or 18.5 μg Cd increase/g lettuce for 50% of diet grown on strongly acidic, sludge-amended soils.

7.1 The Revised "Worst-Case" Cd Transfer Estimate

The true worst-case is more precisely evaluated by the method proposed by Chaney et al. (1987) than is the low sludge Cd case which provides an even higher margin of safety. The next step is to combine the best data on food intakes and on relative increased Cd uptake. Table 5 shows the average adult intake of food groups (averaged over male and female, age groups 14–16 years, 25–30 years, and 60–65 years), based on the published data of Pennington (1983). Dry matter percentage for each food was obtained from Adams (1975). Further, because food groups contained crops with large differences in relative Cd uptake, leafy vegetables, root vegetables, and garden fruits were separated into higher and lower Cd accumulating groups. The new estimated sum of increased Cd intakes, due to consumption of 100% of garden foods (grown on acidic, sludge-amended soil) for the average adult is 2.20 μg Cd day^{-1} when lettuce is increased by 1 mg Cd kg^{-1} dry weight. If one assumes that only 50% of one's garden foods are grown on acidic, sludge-amended soils, dietary Cd increases only 1.10 μg day^{-1} when lettuce Cd is increased by 1 mg kg^{-1} dry weight. This is clearly a substantial reduction in estimated Cd transfer to garden foods. This reduction resulted from changes in the US diet (less potato consumption), change to adult diets to reflect chronic lifetime Cd intakes, and improved estimates of crop and food group Cd uptakes.

This estimated 2.20 μg Cd day^{-1} (100% of garden vegetables) if lettuce is increased by 1 mg Cd kg^{-1} dry weight can be used with the data for increased lettuce Cd when the plateau is reached on sludge-amended soil. The data of Chaney (Fig. 1) were subjected to nonlinear "plateau" regression analysis to estimate the plateau in lettuce Cd for the acidic, heat-treated sludge. The estimated increase above control was 0.93 mg Cd kg^{-1} dry lettuce (upper 95% confidence interval was 1.21 mg Cd kg^{-1} based on nonlinear regression using three replications for 7 crop years) in strongly acidic soil, the worst-case condition, and

0.29 mg kg^{-1} (upper 95% confidence interval was 0.40) in limed soil when low Cd, Fe-rich sludge was applied. This converts to about 1.02 (upper 95% confidence interval = 1.33) μg Cd day^{-1} increased Cd intake (assuming 50% of dietary garden vegetables are grown on strongly acidic, sludge-amended soils). Although earlier estimates suggested only a 30 μg Cd day^{-1} increase could be tolerated, newer data indicate errors were made in these estimates. Average US adult Cd intake is about 15–20 (e.g., 18) μg day^{-1} (Kowal et al. 1979). Fox (1988) noted that the FAO/WHO Potentially Tolerable Weekly Intake for 70 kg adults was 66.7–83.3 (mean 75) μg day^{-1}, not the 57.1–71.4 μg day^{-1} (mean 64) usually quoted. Subtracting 18 from 75 μg Cd day^{-1} leaves 57 μg Cd day^{-1}. However, the oyster-eater data indicate that even greater tolerance exists for individuals consuming Western diets (see above).

Prediction of changes in kidney Cd due to increases in dietary Cd from foods grown in acidic, sludge-amended gardens must also consider effects of nutritional status and nutrients in the garden crops on Cd retention by humans. Research has shown that although crops may have increased Cd concentration, they may not increase kidney Cd, especially when the sludge had low Cd concentration (Chaney et al. 1978a, b; Boyd et al. 1982). Application of low-metal sludges thus constitutes low risk because they allow less Cd uptake by crops, and because the bioavailability of Cd in these crops is low.

In the same way, limiting sludge element concentrations to levels which cannot cause food-chain risk or phytotoxicity is an alternative to Cd or other element application limits (kg ha^{-1}). This is called the "Limit Sludge Quality" or "Good Sludge" approach. Based on the new understandings reviewed in this chapter, it is clear that this approach is much more convenient, allows much more sludge benefit, and greatly reduces the risks from sludge-applied elements in sludge utilization programs. A sludge quality rule, coupled with a waiting period for pathogen demise, and controls on persistent organic compounds in the sludge, allow protection of humans and the environment while recovering the benefits of sludge nutrients and organic matter.

References

Adams CF (1975) Nutritive value of American foods in common units. Agric Handb 456, US Gov Print Off 0100-03184

Baxter JC, Barry B, Johnson DE, Kienholz EW (1982) Heavy metal retention in cattle tissues from ingestion of sewage sludge. J Environ Qual 11:616–620

Beaton GH, Milner J, McGuire V, Feather TE, Little JA (1983) Sources of variance in 24-hour dietary recall data: implications for nutrition study design and interpretation. Carbohydrate sources, vitamins, and minerals. Am J Clin Nutr 37:986–995

Bertrand JE, Lutrick MC, Breland HL, West RL (1980) Effects of dried digested sludge and corn grown on soil treated with liquid digested sludge on performance, carcass quality, and tissue residues in beef cattle. J Anim Sci 50:35–40

Bertrand JE, Lutrick MC, Edds GT, West RL (1981) Metal residues in tissues, animal performance and carcass quality with beef steers grazing *Pensacola bahiagrass* pastures treated with liquid digested sludge. J Anim Sci 53:146–153

Boyd JN, Stoewsand GS, Babish JG, Telford JN, Lisk DJ (1982) Safety evaluation of vegetables cultured on municipal sewage sludge-amended soil. Arch Environ Contamin Toxicol 11:399–405
Bray BJ, Dowdy RH, Goodrich RD, Pamp DE (1985) Trace metal accumulations in tissues of goats fed silage produced on sewage sludge-amended soil. J Environ Qual 14:114–118
Bremner I (1981) Effects of the disposal of copper-rich slurry on the health of grazing animals. In: L'Hermite P, Dehandtschutter J (eds) Copper in animal wastes and sewage sludge. Reidel Publ, Boston, pp 245–260
Bruemmer GW, Gerth J, Herms U (1986) Heavy metal species, mobility and availability in soils. Z Pflanzenernahr Bodenk 149:382–398
Campbell JK, Mills CF (1979) The toxicity of zinc to pregnant sheep. Environ Res 29:1–13
CAST (1980) Effects of sewage sludge on the cadmium and zinc content of crops. Counc Agric Sci Technol Rep 83. Ames, IA, 77 pp
Chaney RL (1980) Health risks associated with toxic metals in municipal sludge. In: Bitton G, Damron BL, Edds GT, Davidson JM (eds) Sludge – health risks of land application. Ann Arbor Sci Publ Inc, Ann Arbor, MI, pp 59–83
Chaney RL (1983) Potential effects of waste constituents on the food chain. In: Parr JF, Marsh PB, Kla JM (eds) Land treatment of hazardous wastes. Noyes Data Corp, Park Ridge, NJ, pp 152–240
Chaney RL (1985) Potential effects of sludge-borne heavy metals and toxic organics on soils, plants, and animals, and related regulatory guidelines. In: Final Rep Worksh International transportation, utilization or disposal of sewage sludge including recommendations. PNSP/85-01. Pan Am Health Org, Washington, DC, Annex 3, Worksh Pap 9, pp 1–56
Chaney RL, Lloyd CA (1979) Adherence of spray-applied liquid digested sewage sludge to tall fescue. J Environ Qual 8:407–411
Chaney RL, Stoewsand GS, Bache CA, Lisk DJ (1978a) Cadmium deposition and hepatic microsomal induction in mice fed lettuce grown on municipal sludge-amended soil. J Agric Food Chem 26:292–994
Chaney RL, Stoewsand GS, Furr AK, Bache CA, Lisk DJ, (1978b) Elemental content of tissues of guinea pigs fed Swiss chard grown on municipal sewage sludge-amended soil. J Agr Food Chem 26:994–997
Chaney RL, Sterrett SB, Morella MC, Lloyd CA (1982) Effect of sludge quality and rate, soil pH, and time on heavy metal residues in leafy vegetables. In: Proc. 5th Annu Madison Conf Appl Res Pract Municipal Ind Waste. Univ Wisconsin-Extension, Madison, Wisconsin, pp 444–458
Chaney RL, Bruins RJF, Baker DE, Korcak RF, Smith JE Jr, Cole DW (1987) Transfer of sludge-applied trace elements to the food-chain. In: Page AL, Logan TJ, Ryan JA (eds) Land application of sludge. Lewis, Chelsa, MI, pp 67–99
Chang AC, Hinesly TD, Bates TE, Doner HE, Dowdy RH, Ryan JA (1987) Effects of long-term sludge application on accumulation of trace elements by crops. In: Page AL, Logan TJ, Ryan JA (eds) Land application of sludge. Lewis, Chelsea, MI, pp 53–66
Chino M, Baba A (1981) The effects of some environmental factors on the partitioning of zinc and cadmium between roots and tops of rice plants. J Plant Nutrit 3:203–214
Corey RB, King LD, Lue-Hing C, Fanning DS, Street JJ, Walker JM (1987) Effects of sludge properties on accumulation of trace elements by crops. In: Page AL, Logan TJ, Ryan JA (eds) Land application of sludge. Lewis, Chelsea, MI, pp 25–51
Dean RB, Suess MJ (eds) (1985) The risk to health of chemicals in sewage sludge applied to land. Waste Manag Res 3:251–278
Decker AM, Chaney RL, Davidson JP, Rumsey TS, Mohanty SB, Hammond RC (1980) Animal performance on pastures topdressed with liquid sewage sludge and sludge compost. In: Proc Nat Conf Municipal and industrial sludge utilization and disposal. Information Transfer, Silver Spring, MD, pp 37–41
EPA (Environmental Protection Agency) (1979a) Criteria for classification of soild waste disposal facilities and practices; Interim final, and proposed regulations. 40 CFR 257. Docket 4004. Fed Reg 44 (179):53438–53468
EPA (Environmental Protection Agency) (1979b) Background document: cumulative cadmium application rates. Criteria for classification of solid wastes disposal facilities. 40 CFR 257. Sept 1979. US-EPA Off Solid Wastes

Evans KJ, Mitchell IG, Salau B (1979) Heavy plant accumulation in soils irrigated by sewage and effect in the plant-animal system. Progr Water Technol 11:339–352
Flanagan PR, McLellan JS, Haist J, Cherian G, Chamberlain MJ, Valberg LS (1978) Increased dietary cadmium absorption in mice and human subjects with iron deficiency. Gastroenterology 74:841–846
Fox MRS (1988) Nutritional factors that may influence bioavailability of cadmium. J Environ Qual 17:175–180
Fox MRS, Tao SH, Stone CL, Fry BE Jr (1984) Effects of zinc, iron, and copper deficiencies on cadmium in tissues of Japanese quail. Environ Health Pers 54:57–65
Fries GF (1982) Potential polychlorinated biphenyl residues in animal products from application of contaminated sewage sludge to land. J Environ Qual 11:14–20
Fries GF, Marrow GS, Snow PA (1982a) Soil ingestion by dairy cattle. J Dairy 65:611–618
Fries GF, Marrow GS, Snow PA (1982b) Soil ingestion by swine as a route of contaminat exposure. Environ Toxicol Chem 1:201–204
Fukushima MA, Iahizaki A, Sakamoto M, Kobayashi (1973) Cadmium concentration in rice eaten by farmers in the Jinzu River Basin. Jpn J Hyg 28:406–415
Grill E, Winnacker E-L, Zenk MH (1985) Phytochelatins: the principle heavy-metal complexing peptides of higher plants. Science 230:674–676
Hallberg L, Bjorn-Rasmussen E, Rossander L, Suwanik R (1977) Iron absorption from Southeast Asian diets. II. Role of various factors that might explain low absorption. Am J Clin Nutrit 30:539–548
Hansen LG, Washko PK, Tuinstra LG, Dorn SB, Hinesly TD (1981) Polychlorinated biphenyl, pesticide, and heavy metal residues in swine foraging on sewage sludge amended soils. J Agric Food Chem 29:1012–1017
Hansen LG, Chaney RL (1984) Environmental and food chain effects of the agricultural use of sewage sludges. Rev Environ Toxicol 1:103–172
Hinesly TD, Hansen LG, Bray DJ, Redborg KE (1985) Transfer of sludge-borne cadmium through plants to chickens. J Agric Food Chem 33:173–180
Hogue DE, Parrish JJ, Foote RH, Stouffer JR, Anderson JL, Stoewsand GS, Telford JN, Bache CA, Gutenmann WH, Lisk DJ (1984) Toxicologic studies with male sheep grazing on municipal sludge-amended soil. J Toxicol Environ Health 14:153–161
Huebers HA, Huebers E, Csiba E, Rummel W, Finch CA (1987) The cadmium effect on iron absorption. Am J Clin Nutr 45:1007–1012
Jarvis SC, Jones LHP, Hopper MJ (1976) Cadmium uptake from solution by plants and its transport from roots to shoots. Plant Soil 44:179–191
Johnson DE, Kienholz EW, Baxter JC, Spanger E, Ward GM (1981) Heavy metal retention in tissues of cattle fed high cadmium sewage sludge. J Anim Sci 52:108–114
Jones SG, Brown KW, Deuel LE, Donnelly KC (1979) Influence of simulated rainfall on the retention of sludge heavy metals by the leaves of forage crops. J Environ Qual 8:69–72
Kienholz E, Ward GM, Johnson DE, Baxter J, Braude G, Stern G (1979) Metropolitan Denver sewage sludge fed to feedlot steers. J Anim Sci 48:735–741
Kowal NE, Johnson DE, Kraemer DF, Pahren HR (1979) Normal levels of cadmium in diet, urine, blood, and tissues of inhabitants of the United States. J Toxicol Environ Health 5:995–1014
Layrisse M, Martinez-Torres C (1971) Food iron absorption: Iron supplementation of food. Progr Haematol 6:137–160
Logan TJ, Chaney RL (1983) Utilization of municipal wastewater and sludge on land-metals. In: Page AL, Gleason TL, Smith JE, Iskander IK, Sommers LE (eds). Utilization of municipal wastewater and sludge on land. Proc. 1983 Worksh, Univ California, Riverside, CA, pp 235–323
McKenzie-Parnell JM, Eynon G (1987) Effect on New Zealand adults consuming large amounts of cadmium in oysters. Trace Subst Environ Health 21:420–430
McKenzie-Parnell JM, Kjellstrom TE, Sharma RP, Robinson MF (1988) Unusually high intake and fecal output of cadmium, and fecal output of other trace elements in New Zealand adults consuming dredge oysters. Environ Res 46:1–14
McLellan JS, Flanagan PR, Chamberlain MJ, Valberg LS (1978) Measurement of dietary cadmium absorption in humans. J Toxicol Environ Health 4:131–138

Newton D, Johnson P, Lally AE, Pentreath RJ, Swift DJ (1984) The uptake by man of cadmium ingested in crab meat. Human Toxicol 3:23–28

Nordberg M, Nuottaniemi I, Cherian MG, Nordberg GF, Kjellstrom T, Garvey JS (1986) Characterization studies on the cadmium-binding proteins from two species of New Zealand oysters. Environ Health Persp 65:57–62

NRC (National Research Council) (1980) Mineral tolerance of domestic animals. Nat Acad Sci, Washington DC, 577 pp

Pedersen B, Eggum BO (1983) The influence of milling on the nutritive value of flour from cereal grains. 4.Rice Qual Plant 33:267–278

Pennington JAT (1983) Revision of the total diet study food lists and diets. J Am Diet Assoc 82:166–173

Pepper IL, Bezdicek DE, Baker AS, Sims JM (1983) Silage corn uptake of sludge-applied zinc and cadmium as affected by soil pH. J Environ Qual 12:270–275

Poole DBR, McGrath D, Fleming GA, Sinnott J (1983) Effects of applying copper-rich pig slurry to grassland. 3. Grazing trials: stocking rate and slurry treatment. Ir J Agric Res 22:1–10

Rahola T, Aaran R-K, Miettinen JK (1973) Retention and elimination of ^{115m}Cd in man. In: Bujdosa E (ed) Health physics problems of internal contamination. Akademiai Kiado, Budapest, pp 213–218

Robinson NJ, Jackson PJ (1986) "Metallothionein-like" metal complexes in angiosperms: their structure and function. Physiol Plant 67:499–506

Romheld V, Marschner H (1986) Evidence for a specific uptake system for iron phytosiderophores in roots of grasses. Plant Physiol 80:175–180

Ryan JA, Pahren HR, Lucas JB (1982) Controlling cadmium in the human food chain: a review and rationale based on health effects. Environ Res 28:251–302

Sanson DW, Hallford DM, Smith GS (1984) Effects of long-term consumption of sewage solids on blood, milk and tissue elemental composition of breeding ewes. J Anim Sci 59:416–424

Sempos CT, Johnson NE, Smith EL, Gilligan C (1985) Effects of intraindividual and interindividual variation in repeated dietary records. Am J Epidemiol 121:120–130

Shaikh ZA, Smith JC (1980) Metabolism of orally ingested cadmium in humans. In: Holmstedt B, Lauwerys R, Mercier M, Roberfroid M (eds) Mechanisms of toxicity and hazard evaluation. Elsevier/North-Holland Biomedical Press, Amsterdam, pp 569–574

Sharma RP, Kjellstrom T, McKenzie JM (1983) Cadmium in blood and urine among smokers and non-smokers with high cadmium intake via food. Toxicology 29;163–171

Smith GS, Hallford DM, Watkins JB III (1985) Toxicological effects of gamma-irradiated sewage solids fed as seven percent of diet to sheep for four years. J Anim Sci 61:931–941

Sugiura Y, Nomoto K (1984) Phytosiderophores. Structures and properties of mugineic acids and their metal complexes. Struc Bond 58:106–135

Yamagata N, Iwashima K, Nagai T (1975) Gastrointestinal absorption of ingested ^{115m}Cd by a man. Bull Inst Publ Health 24:1–6

Sources and Distribution of Pollutant Elements in the Environment, with Particular Reference to Sewage Sludges

P.H.T. Beckett[1]

Summary

The chemistry of pollutant elements in waste water and sludges to a considerable extent depends on their sources. The principal sources and pathways of the eight elements on the EEC Lists I and II are presented. Of the sources that contribute pollutant elements to land, sewage and sewage sludge are the most amenable to control. Their chemistry and assessments of their toxicity are discussed.

1 Introduction

From our present point of view, the most hazardous elements are those on Lists I and II of the EEC Dangerous Substances Directive (EEC 1976): arsenic, cadmium, chromium, copper, lead, mercury, nickel and zinc.

These are also elements for which normal applications of a typical sewage sludge will increase the soil concentration by more than 1 mg kg^{-1} soil *and* by more than 10% of the background soil concentration. In a typical case where digested sludge from a works receiving both domestic and industrial sewage was applied at the maximum level allowed by the guidelines then (1973) current (Chumbley 1971), this list included Ag, As, Ba, Cd, Co, Cr, Cu, Ni, Pb, Sn and Zn (and also Cs and Sm): B, Bi, Hg, Mo and Sb were just outside the limits (Beckett 1978).

We are not concerned here with organo-metallic compounds. So, with the exception of a few very stable complexes of low toxicity, we may assume that the most soluble combinations of these elements are also the most toxic, and that the toxicity of these elements in sludge or waste-water depends on the concentration of their soluble forms, and on the rate at which their insoluble forms of combination may be mobilized.

The following chapter uses data from the United Kingdom to illustrate sources of pollution by the eight pollutant elements on the EEC lists, and the pathways of their dispersion, particularly those which finally bring the elements into the agricultural or natural landscape.

[1]Department of Plant Sciences, University of Oxford, Agricultural Science Building, Parks Road, Oxford OX13PF, Great Britain

2 Sources of Inorganic Pollutants

Most, though not all, of the load of these elements that enter the UK environment is of industrial origin. Table 1a, b estimates the emission from the main groups of sources. Note that the item "municipal waste disposal" includes domestic as well as industrial wastes.

Table 1c indicates the first division of these elements, and Table 1d the percentage of each that proceeds through the public sewers.

Note that there are numerous areas of mining and refining wastes, from mines and smelters now abandoned, with raised concentrations of pollutant elements (Webb et al. 1978), which continue to emit pollutant elements such as dust into river systems (Abdullah and Royle 1972), not included in these statistics.

Table 1. Total UK emission of the pollutant elements and their final destinations (Critchley and Agg 1986)

	As	Cd	Cr	Cu	Hg	Ni	Pb	Zn
a) Contributions to the total emission (%)								
From industry[a]	92	25	61	23	50	27	33	31
Municipal waste disposal	8	67	34	65	31	71	65	64
Domestic	–[b]	1	< 1	4	< 1	< 1	< 1	1
Urban run-off	–	< 1	< 1	< 1	< 1	< 1	< 1	< 1
Agriculture	< 1	5	2	5	13	< 1	< 1	1
Dredging spoil	–	1	3	3	5	1	1	2
b) Total emission (Mg yr^{-1})	2630	754	19 300	15 800	214	14 800	51 700	85 000
c) Total emission received by different environmental compartments (%)								
Atmosphere	13	2	< 1	< 1	21	3	16	3
Agricultural land	< 1	6	3	7	14	1	< 1	2
Landfill[c]	78	83	85	84	50	91	81	88
Rivers	< 1	< 1	1	1	2	1	< 1	< 1
Coastal waters	8	8	11	8	13	4	3	7
d) Percentage which passes through the sewers	1	7	6	6	6	3	2	3

[a] Includes "sewage load not accounted for."
[b] Insufficient data.
[c] Disposals on "other land" (< 1% for all elements) are omitted.

Table 2 lists the main sources of the discharge of these elements into the atmosphere or landfill. The calculations for municipal waste (33 million tons in 1982) assume average concentrations of As 6, Cd 15.5, Cr 200, Cu 315, Hg 2, Ni 317.5, Pb 1025, Zn 1650 mg kg^{-1} dry solids. Critchley and Agg (1986) emphasized that, for lack of information, many of the figures in Table 2 may be underes-

Table 2. Estimated emission of the pollutant elements into the atmosphere (upper figures) and disposal in landfill (lower figures) (in Mg yr^{-1}) (Critchley and Agg, 1986)

	As	Cd[c]	Cr	Cu	Hg	Ni	Pb[c]	Zn
Non-ferrous metal	–[a]	3.7	–	31	–[d]	–	43	109
production	206	0.9(c)	21	399	–	22.2	2150(b)	4090
Iron and steel	9.2	3.3	7.1	18	1.1	2.0	488	1420
production	162	40	5710	309	–	108	3020	10300
Production, consumption	8.7	–	–	–	8.2	–	16(c)	285
disposal of metal products	8.3	21(c)	400	30	43	110	–	150
Fossil fuel consumption;								
including emission from	332	1.4	63.4	15.5	29	477	7430[e]	188
motor vehicles	1480	31(b)	3390	1940	2.4	2820	2250(a)	6000
Cement production	–	0.8	–	–	2.0	–	34	–
	–	21	–	–	–	–	858	–
Municipal waste[b]	0.4	4.5	15	16.5	5.4	9	130	330
	197	504	6540	10310	61	10400	33500	53800
Sewage sludge[b]	0.03	0.2	1.8	1.0	0.6	0.4	1.2	9
	3.6	9.2	284	262	1.6	94	153	624
UK Totals Atmosphere	350	14	87	82	45.6	488	8140	2340
(cf. Pacyna et al. 1984)	(164)	(99)	(1130)	(580)	–	(899)	(10100)	(3490)
Landfill	2060	627	16300	13200	108	13600	41900	74900

[a] Inadequate data.
[b] Atmospheric data relate to gaseous losses during incineration (about 10% of municipal waste is incinerated): landfill data include ash from incineration which, together with furnace ash and fly ash, composes about 10% of the total landfill.
[c] Hutton and Symon's (1986) figures are lower (a), higher (b) or much higher (c) than these: they suggest 6800 and 6000 Mg yr^{-1} lead is emitted to the atmosphere and to landfill respectively from the production, etc. of metal products.
[d] Hutton and Symon (1986) suggest 5 Mg yr^{-1}.
[e] Includes lead emission by motor vehicles: assumes that the 30% of tetraethyl lead not so emitted is retained in the engine and ultimately disposed as scrap material.

timates. Indeed, Pacyna et al. (1984) estimated rather larger atmospheric emission of most elements. Hutton and Symon (1986) presented more detailed information for As, Cd, Hg and Pb, and estimated notably higher emissions of Pb from the metal industries both to the atmosphere and to landfill. Clearly, landfill is the principal recipient of the pollutant elements and obviously it is important and presents many management problems, but here we are mainly concerned with the more diffuse dispersal and disposal of these elements into the whole landscape.

With regard to atmospheric emissions Hutton and Symon (1986) pointed out that data as in Table 2 are overall figures. Their environmental impact will depend very much on whether a large proportion of the emission of an element is from only one or a few sources such as the lead-zinc smelter at Avonmouth (e.g. Burkitt et al. 1972), or from several sources such as the 40-plus municipal incinerators in Britain, or from many sources. It is worth noting that worldwide the ratio of atmospheric emissions from industrial and urban sources such as these to emissions from natural sources are Cd 8.8, Cu 3.0, Ni 1.8, Pb 18, Zn 7.2 (Nriagu 1979). However, wind-blown dust, presumably topsoil, comprised a high proportion of the natural emission. While this is clearly relevant to geochemical cycles, it may not be relevant to discussions on environmental pollution.

Table 3 presents the atmospheric concentrations and total annual atmospheric deposition of the same elements at non-urban sites in Britain.

Table 4 estimates the quantities applied, disposed or deposited on UK agricultural land, i.e. 18.4 million ha of the total land area of 24.1 million ha, of which the arable land has been variously estimated as 4.7–5.1 million ha. Several studies have demonstrated that surface soil horizons in Britain, even when distant from specific industrial pollution, show raised concentrations of the pollutant elements, for example Kinniburgh and Beckett (1983) showed increased topsoil concentrations of Pb, Zn, Mn and Cu (in decreasing order) but not of Mg in normal soils upwind from Oxford and beyond the range of sewage sludge disposal. Even on agricultural land, atmospheric deposition appears to be the principle source of accretions of pollutant elements, and in non-agricultural land it must be the principal current external source. However, many areas of non-agricultural land have been extensively polluted by minor smelters, now closed.

Sewage sludge is also an important source. Table 5 summarizes the principal sources of water-borne wastes containing the pollutant elements and their potential undiluted concentrations. However, Trade Effluent Control limits the concentrations of industrial effluents disposed into the public sewers (Table 6). The permitted levels have gradually been reduced since the mid-1970's (Dakers and King, 1981: Brindley et al. 1982). Also, industrial process waters will be diluted by residential waste-waters. Even so, the influent to a sewage treatment works may show very considerable fluctuations in response to diurnal and weekly cycles of industrial activity (Oliver and Cosgrove 1974; Roberts et al. 1977; Beckett 1980), or furtive discharges from back-street electro-plating works.

Critchley and Agg (1986) assumed an average UK domestic discharge of 130 l waste-water $head^{-1}$ day^{-1}. Imhoff et al. (1971) suggested that perhaps 60% of the

Table 3. Concentrations of the pollutant elements in the atmosphere, and their deposition, both at non-urban sites

	As	Cd	Cr	Cu	Hg	Ni	Pb	Zn
) Total atmospheric mission for UK (Mg y^{-1} rom Table 2: (Pacyna t al. 1984, in parentheses)	350 (164)	14 (99)	87 (1130)	82 (580)	45.6 -	488 (899)	8140 (10 100)	2340 (3490)
) Median and range of oncentrations in the air ear the ground[a] mg kg^{-1} air)	6.5 (2- 19)	< 5 (4–18)	3.6 (1–14)	27 (< 1–55)	0.9 (.05–.3)	5.2 (1–54)	103 (35–380)	194 (64–415)
) Median and range of otal annual deposition at even UK sites[b] g ha^{-1} y^{-1})	> 30 (7.5–55)	-	25 (20–88)	210 (98–480)	< 2 (0.8–18)	92 (35–800)	250 (160–450)	520 (370–1200)
) Median and range of otal annual deposition at ix EHF[c] g ha^{-1} y^{-1})	-	-	10 (5–11)	160 (40–450)	-	36 (18–62)	84 (24–164)	2230 (1530–4340)
) Percent deposition								
n i) Rain	57	-	72	100	-	57	100	91
ii) Dry deposits[d]	43	-	28	0	-	43	0	9

[a] Data from Collafirth (Shetland), Wraymires (Cumbria), Plynlimon (Powys), Trebanos (W. Glam.), Styrrup (Notts), Chilton (Oxon), Leiston (Suffolk coast): Jan.-March 1972 (Peirson et al. 1973).
[b] Same stations, 1972–75: lowest values from Collafirth, Wraymires or Leiston: highest values from Trebanos or Styrrup (Cawse 1980).
[c] Data from experimental husbandry farms, or experimental horticulture stations at Great House (Lancs), Pwllpeiran (Dyfed), Drayton (Worcs), Boxworth (Cambs), Efford (Hants), Lee Valley (Herts) 1967/8–1972 (Wadsworth and Webber 1980).
[d] Data from Chilton (Oxon), a median station from the list for Footnotes a and b, July 1972-June 1973 (Cawse 1980).

Table 4. Total disposals of the pollutant elements onto agricultural land (in Mg yr^{-1}) (Critchley 1983, Critchley and Agg 1986)

	As	Cd	Cr	Cu	Hg	Ni	Pb	Zn
Atmospheric deposition	280	55	466	3700	18	920	4600	14 000
In sewage sludge[a]	2.8	7.5	191	241	1.3	46	155	517
In inorganic fertilizers	10	39	292	62	0.15	47	3.9	431
Other:								
In manure from livestock fed foodstuffs reinforced with trace elements				750				640
In seed dressings					28[b]			

[a] 473 000 Mg (approx. As 6, Cd 16, Cr 400, Cu 510, Hg 3, Ni 97, Pb 330, Zn 1100 Mg kg^{-1} dry solids) were applied to 82 000 ha in 1980/81.
[b] 1977 figure.

Table 5. The principal industrial sources of discharges of the pollutant elements in waste-waters (Patterson, 1979)

Sources		Concentration range ($mg\ l^{-1}$)
As	Ore extraction; manufacture of arsenic trioxide, sulphuric acid, ammonia, insecticides; preservation of wood products	10–1000
Cd	Ceramic and inorganic pigments; electro-plating; batteries; textile printing; manufacture of alloys	4–1000
Cr	Tanning; metal cleaning, treatment and plating, corrosion control additives; pigments and dyes; aluminium anodizing; preservatives	1–3000 (15,–27 000 in plating baths: these may be dumped)
Cu	Metal cleaning, pickling and plating; copper and brass manufactures; primary and secondary copper extraction; manufacture of rayon, copper sulphate, electronics	10–900 (up to 7000 in primary and secondary copper manufactures)
Hg	Chemical manufactures, including chlor-alkali plants; manufacture of chemicals including aceteldehyde; paper mills; water-based paints; fertilizers	< 0.03–20 (once up to 2000)
Ni	Pickling and plating of nickel and nickel alloys; manufacture of steel, motors, aircraft, business machines and tableware	< 1–900
Pb	Manufacture of batteries, ammunition, glass, tetra-ethyl lead, television tubes; pigments, rubber hose; electro-plating	< 1–400
Zn	Zinc-smelting; zinc and brass metal and plating works; steel works; electro-plating; manufacture of wood pulp and newsprint; inorganic pigments	< 1–2000 (2000–35 000 from plating, pickling and borderizing plant)

Table 6. Trade effluent limits ($mg\ l^{-1}$)

	Cd	Cr	Cu	Hg	Ni	Pb	Zn
1[a]	2	25	3	–	10	0.5	15
2	2	5	5	–	4	5	10
3	1	15	5	–	5	10	5
4	2	5	5	1	5	5	10

1[a] Greater Chicago (CAST 1976; Knight and Rinkus 1977);
2 Yorkshire Water Authority (Tench 1977; Thompson and Prior 1984);
3 Thames Water Authority, Mogden Works (Dakers and King 1981);
4 Thames Water Authority, Chiltern Division (Brindley et al. 1982).

load in such discharges will be organic and 40% inorganic (non-combustible); of the latter perhaps 69% will be in solution, 12% in suspension and 19% will be settleable. They also pointed out that the consumption of water per head varies between communities and countries, and the volume of waste-water in proportion. The figure of 130 l $head^{-1}$ day^{-1} may be compared with a median value of 400 l $head^{-1}$ day^{-1} (270–2400 l $head^{-1}$ day^{-1}) to 11 sewage treatment works in West Virginia (Horvath and Koshut 1981). There will also be substantial variations in the proportions of strictly domestic waste-water in sewage from different "residential areas" and Davis and Jacknow (1975) found quite wide variations in the proportions of different elements that were of domestic origin, in sewage from residential areas in New York (Bowery), Pittsburgh and Muncie (Indiana): Cd 38–63%, Cu 36–38%, Pb 10–63%, Zn 16–32%, Ni 13–34%, Cr 3–27%. Brindley et al. (1982) assumed that there must be a substantial domestic component in the copper and zinc reaching a sewage treatment works in an outer suburb of London since their quantities had not decreased during a period of rigorous control of trade effluents, whereas the quantities of Cd, Cr, Ni and Pb had decreased significantly. The contribution per head of pollutant elements to the sewers are correspondingly variable (Table 7).

Table 8a presents analyses for three sets of sewage influents from domestic or non-industrial areas. These may be regarded as the background levels of sludge analyses, onto which the contributions of industrial waste-waters will be superimposed.

Table 9 presents the total quantities of the pollutant elements in water-borne wastes in the United Kingdom. The proportion not received into sewers is mostly discharged into rivers or coastal waters. Part of the load in the sewers is discharged into the sea, more or less untreated, part into the rivers as storm overflow, and the remainder is treated at sewage treatment works. Here, the raw sewage undergoes settlement (*primary treatment*) for up to 24 h, occasionally with added flocculants such as lime, alum or poly-electrolytes. The supernatant (settled sewage or *primary effluent*) may be discharged to the river system, lagooned for natural purification, or passed on for *secondary treatment* by biological (trickle) filtration or activated sludge treatment, followed by further settlement (clarification). *Tertiary treatment* (polishing) with chemical flocculants, or micro-filtration, etc. may follow.

Table 7. Domestic contributions to waste-waters (mg $head^{-1}$ day^{-1})

	Cd	Cr	Cu	Hg	Ni	Pb	Zn
1[a]	5 (2.7)	8.2 (3.2)	45	–	9.1 (5.4)	37 (28)	95 (77)
2	–	4.4	27	–	5.8	12	34
3	0.5	1.2	31	.03	3.5	9.4	40

1[a] Median (and lowest value) of "residential loads" from New York (Bowery), Pittsburgh and Muncie (Ind.) (Davis and Jacknow 1975); 2 Anglian Water Authority (Matthews 1980); 3 calculated from data from Stevenage (Critchley and Agg 1986).

Table 8. Raw sewage and sewage sludge from non-industrial sources in UK

	Cd	Cr	Cu	Hg	Ni	Pb	Zn
a) *Raw Sewage*[a] *(mg l⁻¹)*							
1)	0.004	0.009	0.24	0.0002	0.027	0.072	0.31
2)	0.02	0.03	0.19	–	0.03	0.09	0.24
3)	0.004	0.02	0.06	< 0.001	0.02	0.08	0.1
	-0.01	-0.2	-0.4		-0.2	-0.35	-0.7
b) *Sewage sludge (mg kg⁻¹ dry solids)*							
2)	4.9	48	382	–	31	121	515
4)	7.5	156	332	–	41	270	656

1[a] Stevenage, Herts (Critchley and Agg 1986);
2 Anglia Water Authority, medians of two sets of data (Matthews 1980);
3 R.D. Davis; median values, personal communication;
4 Yorkshire Water Authority (Thompson and Prior 1984).

Table 9. The pollutant elements in water-borne wastes, their treatment in sewage works, and the disposal of the sewage sludge (Mg Yr^{-1}) (Critchley and Agg 1986)

	Cd	Cr	Cu	Hg	Ni	Pb	Zn
Total water-borne wastes	111	2464	2277	33	1233	1692	8402
To sewers	53	1136	1024	14	414	758	2945
To sewage treatment works	30	1122	902	14	374	625	2270
After treatment:							
Effluent	4	401	116	7	186	81	254
Sewage sludge	26	721	786	7	188	544	2013
Sewage sludge to:							
Agricultural land	7.5	191	241	1.3	46	155	517
Other land	1.3	38	36	0.3	10	22	89
Landfill	7.2	212	204	1.5	54	126	503
Incineration	2.2	74	59	0.7	40	28	130
Coastal waters	7.6	206	246	2.9	38	213	774

Table 10 reviews the measures of the efficiency of the treatment. The main purpose of sewage treatment is to recover usable water – perhaps 98–99% of the water in the raw sewage – and to concentrate its organic matter into a small volume of sewage sludge. A more recent goal is also to concentrate the pollutant elements in the sewage into the sludge fraction but this is still a subsidiary purpose. However, for present purposes, the efficiency of sewage treatment may be measured in terms of the fraction of the pollutant elements in the influent sewage that had been removed to produce the final effluent from the works. It is convenient to express this as:

$$\text{Efficiency} = \frac{\text{Influent concentration}}{\text{Effluent concentration}} = R$$

Table 10. The efficiency of sewage treatment processes (expressed as $\frac{\textit{Influent}}{\textit{Effluent}}$ flow or concentration) at removing pollutant elements from returned water. R_1, R_2 and R_3 relate to primary, secondary and tertiary treatment; R_{12} to the combined effect of primary and secondary treatment. (R_{12}) and the figures in parentheses are calculated values of $R_{12} = R_1 \times R_2$

		Cd	Cr	Cu	Hg	Ni	Pb	Zn
(1)[a]	R_1	1.20	1.4	1.9	1.5	–	1.5	1.9
	R_{12}	1.3	3.1	2.7	1.5	–	2.3	2.0
	R_2*		10–20	2–4.8				3.8–20
(2)	R_1	2.3–2.5	2.2–2.5	1.4–1.5	2.5	1.2	2.5–2.9	1.8–3.1
	R_2	2	2.2–2.9	2.4–2.5	>2	1.0	4.6–4.8	1.5–2
	(R_{12})	(4.8)	(5.9)	(5.3)	(>5)	(1.2)	(12.7)	(4.3)
(3)	R_1	1.2	1.2	1.3		1.0	2.2	1.2
	R_2	2.7	2.5	1.7		1.1	2.5	2.3
	(R_{12})	(3.2)	(3)	(2.2)		(1.1)	(5.5)	(2.8)
	R_1	–	4.5–6	3.1–4.4		1–1.6	3.5–5.4	3.3–4.2
(4)	R_1	1.4	1.7	1.7		1.5	2.0	1.5
	R_2	1.5	3.3	2.2		1.8	2.9	2.5
	(R_{12})	(2.1)	(5.6)	(3.7)		(2.7)	(5.8)	(3.8)
(5)	R_1		1.4	1.8		1.3	1.7	1.7
	R'_{2a}		2.4	2.5		1.9	3.4	2.2
	R'_{2b}		3.4	3.6		2.1	7.4	2.7
	R''_2		3.4	4.8		1.4	12.7	2.4
	($R_1 R''_2$)		(4.8)	(8.6)		(1.8)	(21.6)	(4.1)
(6)	R_1	3.6	3.7	3.7		1.3	3.7	3.9
	R_2	2.7	1.5	4.8		2.6	3.7	1.6
	(R_{12})	(9.7)	(5.6)	(17.8)		(3.4)	(13.7)	(6.2)
(7)	R_1	1.3	1.5	1.3		1.3	1.3	1.4
	R_{12}	1.6	2.1	3.1		2.3	2.0	2.0
(8)	R_2	1.6	2.3	1.5	3.2	1.3	2.3	2.9
	R_2,R_{3a}	2.2	4.4	1.6	3.9	1.5	2.9	4.5
	R_2,R_{3b}	1.6	3.1	1.5	3.8	1.5	2.3	2.9
	R_2,R_{3b},R_{3a}	1.6	2.6	2.5	5	2.7	2.3	10.3
	R_2,R_{3b},R_{3a},R_4	1.6	2.6	3.0	7	4.0	2.3	10.3

[a](1) US Midwest. R_2* denotes a pilot-activated sludge treatment (Brown et al. 1973); (2) Ontario (Oliver and Cosgrove 1974); (3) Pittsburgh and Muncie (Ind.) (Davis and Jacknow 1975); (4) Zurich (Roberts et al. 1977); (5) Manchester and Salford; the data also showed that even primary settlement was aided by biological flocculation, and was significantly furthered by additions of flocculants: R′ denotes biological (trickle) filtration, R_{2a} denotes settled trickle effluent, and R_{2b} filtered trickle effluent: R''_2 denotes activated sludge treatment (Stones 1977); (6) Oxford; the authors reported that there was very variable efficiency at nickel removal in both stages, and for Cd and Pb in the activated sludge treatment (Lester et al. 1979; Stoveland et al. 1979); (7) West Virginia; wide variations in the efficiency of removing Cu or Pb (Horvath and Koshut 1981); (8) Dallas; activated sludge treatment (R_2), followed by filtration (R_{3a}) or by addition of flocculants (R_{3b}), or both (R_{3a},R_{3b}), and also by activated carbon treatment (R_{3a},R_{3b},R_4) (Esmond and Petrasek 1974).

Defined in this way the combined efficiency of successive primary, secondary and tertiary treatments is simply the product of their separate efficiencies, i.e. $R_{123} = R_1 \times R_2 \times R_3$.

The efficiencies reported are only crude estimates, and the calculated values of $R_{12} = R_1 \times R_2$ combine the errors of both estimates. Nevertheless, there is a clear pattern in the results, summarized in the median values of R_{12}:

Pb 5.8, Cr 4.8, Zn 3.8, Cu 3.7, Cd 2.8, Ni 2.1.

This sequence may be compared with the corresponding ratios of

$$\frac{\text{Total UK sewage influent}}{\text{Total UK sewage effluent}}$$

from Table 9, i.e. Zn 8.9, Cu 7.8, Pb 7.7, Cd 7.5, Cr 2.8, Ni 2.0.

Clearly, sewage treatment is notably less successful at extracting Cd or Ni from the returned water. Several authors (e.g. Brown et al. 1973; Oliver and Cosgrove 1974; Lester et al. 1979) have pointed out that even secondary treatments are more successful at removing fine particulate material than elements in solution. The effects of filtration in Table 10 (e.g. R^1_{2b} and R_{3a}), and filtration after adding a chemical flocculant, are consistent with this finding.

Table 11 shows how the ratio of the dissolve/total concentrations increases through primary settlement and activated sludge treatment followed by clarification. It also demonstrates that the efficiency of removal is greatest where the concentration is greatest.

Table 11. Purification by primary and secondary treatment (in mg l^{-1}) (Oliver and Cosgrove 1974)

	Cd	Cr	Cu	Hg	Ni	Pb	Zn
Total concentrations							
Influent sewage	0.006	0.29	0.31	0.007	0.33	0.23	2.40
Primary effluent	0.003	0.013	0.21	0.003	0.28	0.09	1.13
Secondary effluent	0.001	0.06	0.08	<0.001	0.27	0.015	0.56
Ratio of dissolved/total concentrations							
Influent	<0.2	0.07	0.55	0.09	0.67	0.05	0.24
Primary effluent	<0.35	0.08	0.76	0.1	0.79	0.13	0.51
Secondary effluent	–	–	0.88	–	0.81	0.4	0.71

Table 12 presents typical compositions of final effluent, together with the permissible upper limits, and the composition of a "clean" stream.

The sludge precipitates, i.e. primary sludge, humus sludge and surplus activated sludge, which contain the removed pollutant elements are either dumped, incinerated or treated further before disposal as digested sewage sludge. Horvath and Koshut (1981) reported a median value of 180 g dry solids $head^{-1}$ day^{-1} (65–700) from 11 treatment works in Virginia.

Table 12. The composition of the final effluents after sewage treatment (mg l^{-1})

	Cd	Cr	Cu	Hg	Ni	Pb	Zn
Permissible upper limits							
(1)[a]	1.5		1.0		1.0	0.1	1.0
(2)	0.2	0.5	0.5		0.5	0.5	2.0
(3)	0.1	2	0.5		1.0	0.5	2.0
Reported values							
(4)	0.02	0.08	0.10		0.08	0.06	0.53
(5)	0.001	0.06	0.08	< 0.001	0.27	0.015	0.56
(6)		0.04	0.06		0.09	0.27	0.27
(3)	0.003	0.03	0.06			0.05	0.25
(7)	0.08–0.1	0.1			0.07–0.08	0.03–0.1	0.3–1.5
(8)		0.02	0.005		0.03	0.01	0.04
(9)	< 0.001	0.02	0.007		0.01	0.008	0.08
(10)	0.003	0.03	0.06	0.007	0.03	0.07	0.20
"Clean" stream water (in μg l^{-1})							
(11)	0.41		0.66		0.50	0.70	11.0

[a](1) Greater Chicago (IEPA; Knight and Rinkus 1977); (2) Yorkshire Water Authority (Tench 1977); (3) Switzerland (Roberts et al. 1977); (4) Pittsburg (Davis and Jacknow 1975); (5) Canada (Oliver and Cosgrove 1974); (6) Muncie, Ind. (Davis and Jacknow 1975); (7) UK (Stones 1977); (8) Greater Chicago (Knight and Rinkus 1977); (9) Oxford, UK (Lester et al. 1979; Stoveland et al. 1979); (10) West Virginia (Morvath and Koshut 1981); (11) minimum levels in streams of West Wales, free of domestic or industrial pollution (Abdullah and Royle 1972).

The Institute for Water Pollution Control (IWPC 1975) reported that the sewage sludges being disposed onto land at that time (probably some of them onto sacrificial land) had median contents of:

Cd 40, Cr 330, Cu 560, Ni 100, Pb 670, and Zn 1650 mg kg^{-1} dry solids. These all contain more or less industrial discharges. Table 8b presents the compositions of sludges from works receiving only domestic or non-industrial sewage: these probably represent the minimum currently attainable loadings. Tables 9 and 13 outline the disposal of sludge in the United Kingdom and in the areas of two Water Authorities.

The preceding sections have discussed the sources, and subsequent divisions, of pollutant elements in sewage sludge. This is very relevant to discussions on how sewage may best be treated and how sewage sludge may best be disposed to minimize environmental pollution. The following sections discuss changes in the forms of combination of the pollutant elements, and how the toxicity of sludge or sludge-treated soil may be assessed.

Table 13. Disposal of sewage sludge in the United Kingdom (Matthews 1979, 1980; Thompson and Prior 1984)

UK disposal –	Agricultural land:	arable 26%, grazing 16%, reclamation 5%, horticulture and allotments 2%
	Landfill	24%
	Sea	23%
	Incineration	3%
Anglia Water Authority –	Agricultural land:	liquid-digested sludge 22–24%, chemically conditioned cake 39–44%, primary, lagooned, dried, cold-digested 12–18%
	Non-agricultural land	5 -17%
	Sea	10–11%
Yorkshire Water Authority –	Liquid sludge:	to land 33%, to sea 15%, lagooned 13%
	Cake	to land 9%, to tip 15%
	Ash	to tip 10%
	Sold	5%

3 Transformations Among the Pollutant Elements

On the combination forms of the pollutant elements in the atmospheric deposition that forms a substantial part of their accretion in the landscape, little information is as yet available. Even in the rural area of Wraymires in NW England (Peirson et al. 1973) the As, Cr, Pb and Zn deposits could be shown to be of industrial (or urban) origin. Their small particle size was consistent with their having been formed by the combustion of fossil fuels and municipal waste, which Table 2 suggests must have provided a considerable part of them. If so, the high temperature compounds formed during combustion are likely to revert to other compounds in the soil.

Among the industrial wastes that enter the sewers we should recognize that part of the potentially toxic elements is present in relatively inert forms, such as metal fragments, some oxide or sulphide pigments or fillers, and some soil minerals, that pass through the sewers and sewage treatment works relatively unchanged. It is possible that a substantial proportion of the tin in the sewage system may be of this kind. Most of the remainder of the pollutant elements in industrial waste-waters may be in soluble or fine particulate forms. Most of the former are likely to be precipitated as carbonates or absorbed on/in freshly precipitated iron or manganese oxides on being mixed with percolating water and domestic sewage.

The crystallization of iron oxides is strongly inhibited by organic anions, so there will be extensive formation of colloidal iron-humus colloids with a substantial capacity for complexing or adsorbing pollutant cations and anions.

These are likely to be removed by normal sewage treatment. However, small stable organic ligands in the waste-waters, resistant to microbial attack, might well increase the proportion of pollutant cations in the final effluent.

In the lower reaches of the sewers, anaerobic conditions may reduce some iron (III) precipitates and re-mobilize any pollutant elements they contain, thus generating sulphides. There will be some precipitation of sulphides and carbonates in the sewers and as a result of continuing biological processes during primary sedimentation. The primary sludge becomes increasingly anaerobic during storage.

Conversely, surplus activated sludge and humus sludge from biological filtration are aerobic. Both are enriched by metabolites from active microbial activity, which have substantial capacities for complexing pollutant cations or adsorbing finely particulate colloids, and which also provide a readily degradable substrate when aerobic sludges are mixed with the primary sludge for lagooning, cold digestion or mesophilic anaerobic digestion. So far these anaerobic processes have been preferred for sanitary and other reasons, and because of low winter temperatures. The main form of sludge disposed onto farmland has been anaerobically digested sludge. However, there are problems in maintaining the digesters, thus aerobic digestion may be preferred.

Digestion, and particularly mesophilic anaerobic digestion, leads to radical transformations in the combination forms of pollutant elements. Decomposition of the high-energy, non-resistant, organic substances from the sewage, and from sludges from secondary treatments, raises the pH and produces substantial quantities of dissolved CO_2 and carbonate ions. It also releases phosphate, sulphate reduced to sulphide, and ammonium, from organic residues. Iron, chromium and manganese are converted to their reduced and more soluble ions. As a result, there is extensive precipitation of calcite, the calcium phosphates (starting with brushite and ending with apatite) and possibly magnesium ammonium phosphate and iron (II) phosphates. Few of these are pure, and the precipitate phase appears to contain some or all of Ca-Fe(II)-carbonates and a range of Ca-Fe(II)-(Al-Mn)-carbonate-hydroxide-phosphates. The latter group may contain significant proportions of pollutant cations. At the normal digester pH, equilibrium between iron (II) carbonate and hydrated iron (III) oxides need not always incline to the former. Possibly more significantly, a wide range of identifiable sulphides of Zn, Cu, Ni, Ba and Pb are formed, dominated by iron (II) sulphides. Groups of these are mixed at the micro-crystalline level.

Even so, very substantial proportions of the pollutant elements and iron remain in association (adsorbed or as co-precipitates) with colloidal organic matter presumably consisting of resistant hetero-polymers from microbial residues, or metabolites from active anaerobes. It is likely that the inorganic precipitates control the composition of the digester solution, but the colloid phase may provide its principal buffer. The high ionic strength, the predominance of organic ligands of unknown and probably varying capacities for complexation, and the multiplicity of imperfectly crystalline, inorganic precipitates, make exact calculations of digester equilibria difficult. Some attempts have been made. It is important to establish the proportions of the combinations of pollutant elements which may be expected to survive aeration of the digested sludge, and the proportions which will undergo some degree of mobilization and rearrangement on oxidation.

When an anaerobic sludge is disposed onto agricultural land, the change in its physico-chemical environment is drastic. There will be a rapid rise in redox potential and a surge in microbial activity as the demise of anaerobic bacteria provides a temporary bonus of microbial substrate. Iron (II) compounds appear to oxidize rapidly, with the precipitation of active iron (III) and manganese (IV) oxides according to the prevailing pH. The stability of the carbonates and phosphate precipitates will depend upon the soil pH, and is likely to be least for the precipitates with the highest proportions of substituted iron (II) cations. In the long term, phosphates will revert to combination forms determined by soil pH, while chlorides and temporary excesses of nitrate and soluble organic matter will be lost by leaching (Campbell and Beckett 1988), until the controlling anions become the normal chloride, bicarbonate, sulphate and nitrate.

The oxidation of the sulphide precipitates is likely to proceed along pathways already elucidated in the case of acid sulphate soils. There is some evidence that pollutant cations may escape from the precipitates upon oxidation of the sulphide, but before the oxidation of iron (II) and its precipitation as hydrated iron (III) oxides. This does not preclude their subsequent re-adsorption and possible occlusion within new soil precipitates of iron (III) oxides, for which there is also some evidence.

The final form of the new equilibrium or steady state established will be complex, and certainly not yet predictable. It is doubtful whether it will be greatly elucidated by studies involving sequential extractants (Beckett 1989) until there has been considerably more work done to establish which of the extractants mobilize definable and mutually exclusive pools of minor elements, without the rapid re-adsorption of the dissolved forms into other combination forms.

A priori, however, it seems clear that as long as the concentration of the soluble and available forms of a pollutant element in the soil solution of a sludged soil is controlled mainly by microbial metabolites, the availability or toxicity of that element is unlikely to show any consistent reversion to less soluble or toxic forms. On the other hand, once it is clear that the availability of a pollutant element is controlled by inorganic physico-chemical processes, then it may be assumed that it will gradually change, however slowly, to the form appropriate to the particular soil environment. For example, Mattigod et al. (1986) found that zinc and lead in the controversial mining wastes at Shipham (Somerset) had formed crystalline carbonates, and inferred that Cu and Cd may have been present as solid solutions in the zinc carbonates. This may explain why they are substantially less toxic (Davies and Ginnever 1979) than the current procedures for environmental monitoring had indicated. The concept of change through a sequence of progressively less soluble/available forms is familiar to students of soil phosphate.

4 Assessments of the Toxicity of Pollutant Elements in Soils, or Sludge-Treated Soils

The chemistry of minor elements in soils is complex at best, and their chemistry in soils recently treated with sludge and still undergoing relatively rapid transformations is even more complex, and assessments of the deficiency or toxicity of an element by soil or sludge analysis are rarely clear.

While it is probably necessary to use extractants such as acetic acid, EDTA (MAFF 1981) or DTPA (e.g. Lindsay and Norvell 1978; Korcak and Fanning 1978; Baker and Amacher 1981) problems have been reported for sludge-treated soils. Beckett et al. (1983) have shown that the availability to plants of Cu and Zn from a sludge-treated soil and their extractability to conventional reagents both changed as soil and sludge interacted, but not always in the same direction.

In view of these problems, and the need to recalibrate the interpretation of measurements with extractants for different situations, it may be easier to use plants to measure toxicity. Young plants are very sensitive to changes in availability, and their tissue is nearly all of the same physiological age. Beckett and Davis (1977) and Davis and Beckett (1978) were able to show for young plants of several species and particularly for young barley that the relation between growth, or the depression of growth by toxicity, and the concentration of a potentially toxic element in the plant tissue was remarkably independent of external factors. Over the toxic range, growth decreased in proportion to increases in the logarithm of tissue concentration. In contrast, the relation between growth and the concentration or availability of a toxic element around the plant roots proved to be very sensitive to other factors. It does seem relatively easy to define a critical tissue concentration for each element and then to ascertain (by analyzing young five-leaf, 2–3-week-old barley test plants) whether plants on a particular site are likely to suffer from toxicity, and roughly how far they find themselves above or below the onset of toxicity. Davis et al. (1978) presented a preliminary list of the critical tissue concentrations of 20 potentially toxic elements, more or less consistent with other published work. Macnicol and Beckett (1985) presented an exhaustive survey of critical tissue concentrations extracted from the literature.

5 Conclusion

This chapter has attempted to trace the sources and pathways of the pollutant elements that accrete on the agricultural and natural landscapes. More detailed information would help to clarify their chemistry. It discusses, with more confidence, the pathway that is most amenable to control, namely the production and disposal of sewage sludge. The question of whether or not elements added to the soil, along any of these paths, gradually revert to less soluble and less toxic forms is crucial.

References

Abdullah MI, Royle LG (1972) Heavy metal content of some rivers and lakes in Wales. Nature (London) 238:329-330

Baker DE, Amacher MC (1981) The development and interpretation of a diagnostic soil-testing program. Penn State Univ Exp Stn Bull 826

Beckett PHT (1978) An all-element analysis of digested sewage sludge. Water Pollut Control, 77:539-547

Beckett PHT (1980) The statistical distribution of sewage and sludge analyses. Environ Pollut 1:22-35

Beckett PHT (1989) The use of extractants in studies on trace metals in soils, sewage sludges, and sludge-wasted soils. Adv Soil Sci 9:143-176

Beckett PHT, Davis RD (1977) Upper critical levels of toxic elements in plants. New Phytol 79:95-106

Beckett PHT, Warr E, Davis RD (1983) Cu and Zn in soils treated with sewage sludge: their "extractability" to reagents compared with their "availability" to plants. Plant Soil 70:3-14

Brindley P, Carter DC, Linsmith MJ (1982) Industrial effluent control experience. Effluent Water Treat J 22:303-310

Brown HG, Hensley CP, McKinney GL, Robinsion JL (1973) Efficiency of heavy metals removal in municipal sewage treatment plants. Environ Lett 5:103-114

Burkitt A, Lester P, Nicklus G (1972) Distribution of heavy metals in the vicinity of an industrial complex. Nature (London) 238:327-328

Campbell DJ, Beckett PHT (1988) Soil solutions extracted from soil treated with digested sewage sludge. J Soil Sci 39:283-298

CAST (Council for Agricultural Science and Technology) (1976) Application of sewage sludge to cropland: appraisal of potential hazards of the heavy metals to plant and animals. Counc Agric Sci Technol Rep 64

Cawse PA (1980) Deposition of trace elements from the atmosphere in the UK. In: Inorganic pollution and agriculture. Reference book 326. MAFF, London, pp 22-46

Chumbley CG (1971) Permissible levels of toxic metals in sewage used on agricultural land. ADAS Advis Pap 10. MAFF, London

Critchley RF (1983) An assessment of trace metal inputs and pathways to the marine and terrestrial environments. Proc Int Conf on heavy metals in the environment, Sept 1983. CEP, Edinburgh, pp 1108-1111

Critchley RF, Agg AR (1986) Sources and pathways of trace metals in the United Kingdom. Rep ER 822-M. WRC Environment, Medmenham, UK

Dakers JL, King RP (1981) Control of toxic metals at sewage treatment works. Water Sci Technol 14:53-59

Davies BE, Ginnever RC (1979) Trace metal contamination of soils and vegetables in Shipham, Somerset. J Agric Sci (Cambs) 93:753-756

Davis JA, Jacknow J (1975) Heavy metals in waste-water in three urban areas. J Water Pollut Control Fed 47:2292-2297

Davis RD, Beckett PHT (1978) Upper critical levels of toxic elements in plants. II. Critical levels of copper in young barley, wheat, rape, lettuce and ryegrass, and of nickel and zinc in young barley and ryegrass. New Phytol 80:23-32

Davis RD, Beckett PHT, Wollan E (1978) Critical levels of twenty potentially toxic elements in young spring barley. Plant Soil 49:395-408

EEC (European Economic Community) (1976) Directive on the discharge of dangerous substances. On pollution caused by certain dangerous substances discharged into the aquatic environment of the community. Counc Eur Communities, 76/464/EEC-OJL 129

Esmond SE, Petrasek AC (1974) Trace metal removal. Ind Water Eng 1:14-17

Horvath DJ, Koshut RA (1981) Properties of several elements found in sewage effluent and sludge from several municipalities in West Virginia. J Environ Qual 10:491-497

Hutton M, Symon C (1986) The quantities of cadmium, lead, mercury and arsenic entering the UK environment from human activities. Sci Total Environ 5:129-150

Imhoff K, Müller WJ, Thistlethwayte DKB (1971) Disposal of sewage and other water-borne wastes, 2nd edn Butterworths, London

IWPC (Institute of Water Pollution Control) (1975) Rep Sub-Committ Disposal of sewage sludge to land. Water Pollut Control 74:476–480
Kinniburgh DG, Beckett PHT (1983) Geochemical mapping in Oxfordshire: a comparison of stream sediment and soil mapping. J Soil Sci 34:183–203
Knight EW, Rinkus RR (1977) Effluent standards for treatment plants and storm water overflows. Progr Water Technol 8:17–26
Korcak RF, Fanning DS (1978) Extractability of Cd, Cu, Ni and Zn by double acid vs. DTPA and plant content at excessive soil levels. J Environ Qual 7:506–512
Lester JN, Harrison RM, Perry R (1979) The balance of heavy metals through a sewage treatment works. I. Lead, cadmium and copper. Sci Total Environ 12:13–23
Lindsay WL, Norvell WA (1978) Development of a DTPA soil test for Zn, Fe, Mn and Cu. Soil Sci Soc Am J 42:421–428
Macnicol RD, Beckett PHT (1985) Critical tissue concentrations of potentially toxic elements. Plant Soil 85:107–129
MAFF (1981) The analysis of agricultural materials. Reference book 427. MAFF, London
Matthews PJ (1979) Sewage sludge disposal – a service to agriculture. Effluent Water Treat J 19:464–471
Matthews PJ (1980) Sewage sludge utilisation – management as part of water quality control. Environ Technol Lett 1:65–67
Mattigod SV, Page AL, Thornton I (1986) Identification of some trace metal minerals in a mine-waste contaminated soil. Soil Sci Soc Am J 50:254–258
Nriagu JO (1979) Global inventory of natural and anthropogenic emissions of trace metals to the atmosphere. Nature (London) 279:409–411
O'Donnel AR, Mance G (1984) Quantities of some dangerous substances discharged to British Coastal Waters. Mar Pollut Bull 15:284–288
Oliver BG, Cosgrove EG (1974) The efficiency of heavy metal removal by a conventional activated sludge treatment plant. Water Res 8:869–874
Pacyna JM, Semb A, Hanssen JE (1984) Emission and long-range transport of trace elements in Europe. Tellus 36B:163–178
Patterson JW (1979) Industry sources and control of heavy metals in the environment.Proc Int Conf on: Management and control of heavy metals in the environment. London, CEP Edinburgh, pp 617–625
Peirson DH, Cawse PA, Salmon L, Cambray RS (1973) Trace elements in the atmospheric environment. Nature (London) 241:252–256
Roberts P, Hegi HR, Weber A, Krähenbühl HR (1977) Metals in municipal waste-water and their elimination in sewage treatment. Progr Water Technol 8:301–306
Stones T (1977) Fate of metals during sewage treatment. Effluent Water Treat J 17:653–655
Stoveland S, Astruc M, Lester JN, Perry R (1979) The balance of heavy metals through a sewage treatment works. II. Chromium, nickel and zinc. Sci Total Environ 12:25–34
Tench HB (1977) Effluent standards with special reference to the River Don system. Progr Water Technol 8:41–47
Thompson KC, Prior BF (1984) Some problems for a water authority caused by toxic metals in trade effluent. Effluent Water Treat J 24:196–200
Wadsworth GA, Webber J (1980) Deposition of minerals and trace elements in rainfall. In: Inorganic pollution and agriculture. Reference book 326, MAFF, London, pp 47–55
Webb JS, Thornton I, Thompson M, Howarth RJ, Lowenstein PL (1978) The Wolfson geochemical atlas of England and Wales. Oxford Univ Press

Part D Case Studies

Introductory Comments. The two chapters in Part D discuss two different field-scale case studies. Both are related to underground water pollution, stemming from vertical flow of water and chemicals through the vadose zone. Shevah and Waldman discussed nitrogen and irrigation management practices used in Israel to increase fertilizer utilization efficiency. The enhanced nitrogen uptake efficiency results in increased plant dry matter production, higher economical yield, and greater water consumption, which reduces the amount of water available for leaching. This, and the smaller fraction of N-fertilizer left in the soil, reduce groundwater contamination caused by mineral N in the soil.

Gvirtzman and Magaritz discussed the use of atmospheric tritium measurements in groundwater as a means for analyzing the travel velocity of water and pollutants in the subsoil and the mechanism of transport and mixing. Their analysis is based on the fact that the tritium level in rainwater has increased since the first nuclear tests were carried out in the atmosphere, hence there is a distinct difference in tritium concentration between rainwater and irrigation water.

J. Goldshmid

Advances in Management of Fertilizer Application and the Resulting Effect on the Pollutant Load in Soil

Y. Shevah[1] and M. Waldman[2]

1 Introduction

Fertilizers have long been seen as an important means of producing the food needed by the population as they replace the nutrients removed by crops and maintain soil fertility. However, the environmental implications of this cause growing concern as evidence increases that application of agrochemicals and fertilizers is reducing the quality of water, particularly in terms of the nitrate content in drinking water. In recent years a significant upward trend in the nitrate content has been recorded, especially in groundwater resources underlying arable land (Ronen et al. 1984). Moreover, recent reports have indicated that the concentration within aquifers can be expected to intensify as a result of the continued accumulation of nitrogen in the soil profile (Jollans 1985, Ronen and Magaritz 1985).

The various sources and pathways of nitrogen through the soil which finally lead to the contamination of groundwater with nitrate were clearly described by Conrad (1986). Fertilizer application is only one source of nitrogen contamination of groundwater. Other major sources are sewage and solid waste, rainfall, and biological nitrogen fixation (Waldman and Shevah 1984). Nevertheless, pollution of water resources has become an emotive issue which has attracted much public awareness.

Although there is no conclusive evidence that high nitrate consumption is positively linked with the blue baby syndrome and gastric cancer (Jollans 1985, Conrad 1986), it is generally agreed that some limit should be applied on nitrate content in water. In accordance, most countries have already passed standards for nitrate content in drinking water and have adopted national policies and regulations to control it.

The regulations usually focus on corrective or compensatory measures in order to meet official standards. The European Commission favours the "polluter pays" principle, whereby the cost of treating pollution is placed on the identified polluter. Mandatory regulations and procedural approaches, such as taxes on nitrogen fertilization, have also been introduced recently in Sweden (Conrad

[1]Tahal Consulting Engineers LTD., Tel-Aviv, Israel
[2]Ministry of Science and Technology, P.O.B. 18195, Jerusalem 91181, Israel

1986). Similarly, the Dutch Government has passed a legislation which enables it to impose restrictions on fertilizer use (Nitrogen 1986). Other corrective measures include: abandonment of contaminated sources; development of new water resources; the treatment of water by denitrification; blending high nitrate water with water of low nitrate content; long-term storage; and ion exchange processes (Jollans 1985).

The need for intensive agriculture requiring large quantities of fertilizer will increase over time. Therefore, in addition to administrative measures, which have been suggested to reduce the amounts of nitrate in run-off, emphasis should be placed on adopting appropriate agrotechniques as a way of increasing the efficiency of nitrogen application. In this way it may well be possible to achieve the necessary environmental improvement without penalizing the farming community.

This chapter analyzes the extent to which such objectives have been achieved in Israel through advances made in the management of fertilizers and consequent improvements in nitrogen application.

2 Methodology

Data on fertilizer use obtained from the Israel Bureau of Statistics, the Ministry of Agriculture, and other sources for 1984 were compiled to produce a partial nitrogen balance. The methodology used was that applied by Saliternick et al. (1972), who estimated a nitrogen balance for Israel's intensively cultivated Coastal Plain Region and the country as a whole in 1970 by taking into account nitrogen applied to crops and nitrogen uptake by harvested crops to give nitrogen residues in soils. The results for 1984 and 1970 were then compared in order to assess the improvement in fertilizer use efficiency within the intervening 15-year period.

3 Results

3.1 Nitrogen Fertilizer Use

Total Use and Types of Fertilizers. Total use of nitrogenous fertilizers in Israel has grown from about 10 000 tons in the 1950's to about 44 000 tons in 1984. Simple compounds such as ammonium sulphate (21% N) and ammonium nitrate (18–21% N) have increasingly been replaced with new ready-mixed NPK fertilizers prepared in solid and liquid forms to suit specific requirements. The use of fertilizers in solution, such as liquid ammonia and urean 32 solution, accounts for about 10 to 20% of the total (Fig. 1).

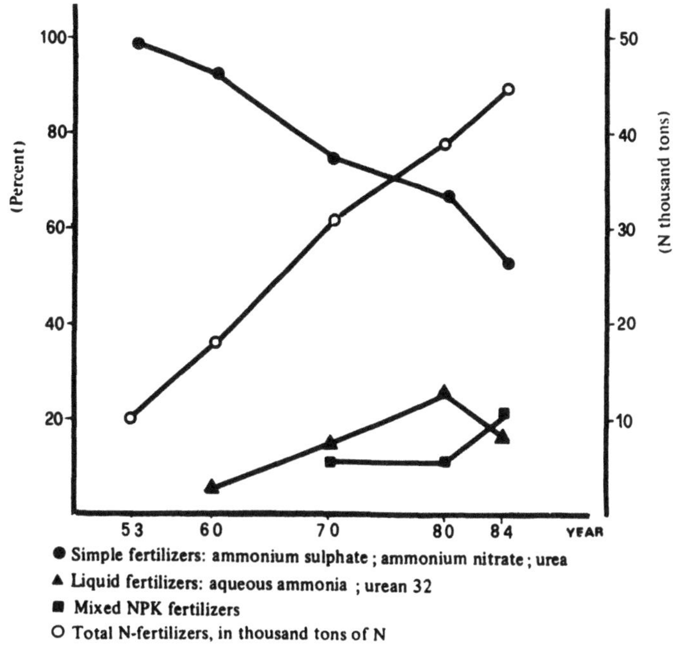

Fig. 1. Total purchases and types of N-fertilizer used, 1953–1984. *Source*: Compiled from Statistical Abstract of Israel (Israel Central Bureau of Statistics 1985)

Expenditure. The general increase in nitrogenous fertilizer usage has been accompanied by a decline in the expenditure on fertilizers relative to that on other purchased inputs such as energy, pesticides, and water. The expenditure on fertilizer has declined from about 8% of total expenditure in the 1950's to about 4% in the 1980's (Fig. 2). The same trends are observed when relating the expenditure on fertilizers to the total value of agricultural production (Fig. 3). Today the expenditure on fertilizer is only 2% of the total value of agricultural production, as compared to about 4% on energy and 5–6% on agrochemicals (pesticides and fungicides).

Cost. The transition from the traditional, simple and stable fertilizers to new, ready-mixed NPK fertilizers has not been accompanied by an increase in purchase and application costs. A detailed analysis of farmgate prices and application costs for five different formulas (Marom 1986) indicates that simple fertilizers cost US$ 150–170 ha^{-1}, as compared to US$ 160–166 ha^{-1} for new, ready-mixed NPK fertilizers (Table 1).

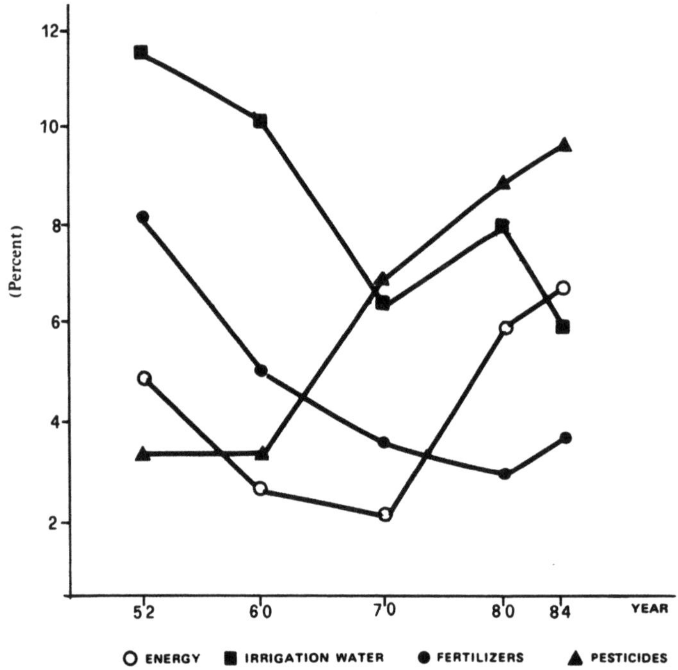

Fig. 2. Expenditure on fertilizer, chemicals, energy, and water as percentage of total purchased inputs in agriculture, 1952–1984. *Source*: Compiled from Statistical Abstract of Israel (Israel Central Bureau of Statistics 1985)

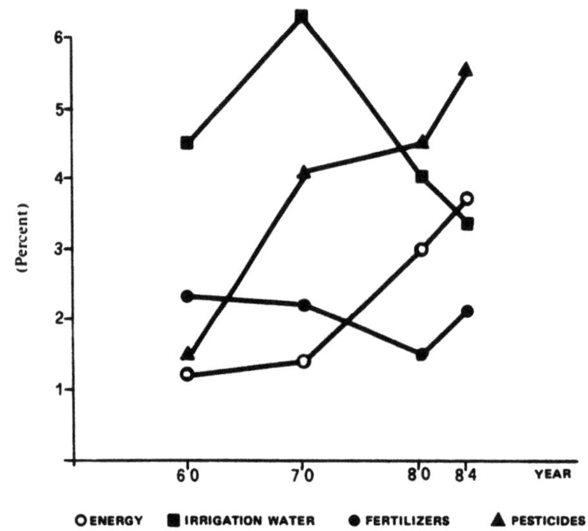

Fig. 3. Expenditure on fertilizer, chemicals, energy, and water in relation to total agricultural production value, 1960–1984. *Source*: Compiled from Statistical Abstract of Israel (Israel Central Bureau of Statistics 1985)

Table 1. Cost of mixed NPK fertilizers compared to simple fertilizers[a] (US$ ha^{-1})

	Simple fertilizers			Mixed NPK	
	A	B	C	D	E
Farmgate price	103	124	130	128	130
Application cost	47	43	40	32	36
Total	150	167	170	160	166

Source: Compiled from D. Marom (1986).
[a] A = Ammonium sulphate; B = aqueous ammonia; C = superphosphate; D = poly'on 16–8–0; E = gar'on 13–7–0.

3.2 Nitrogen Application Rate

The amount of N fertilizers applied to various crops is shown in Table 2 as a weighted average. These figures indicate a continuing increase in the average level of application of N fertilizers between 1970 and 1984, from 116 kg N ha^{-1} to 156 kg N ha^{-1}. However, this increased rate of application has been accompanied by a remarkable increase in average yields (Table 2 and Fig. 4).

Table 2. Nitrogen content, nitrogen applied, and yields for major crops, 1970 and 1984

		1970		1984	
	N-content (%)	N-applied (kg ha^{-1})	Yield (ton ha^{-1})	N-applied (kg ha^{-1})	Yield (ton ha^{-1})
Fruit trees					
Citrus	0.16	200	32	170	45
Deciduous	0.05	150	25	120	42
Tropical	0.10	230	9	180	12
Field crops					
Cotton	2.4	110	3.3	130	4.5
Veg.	0.3	180	33	230	72
Others	0.3	300	45	320	75
Non-irrigated					
Grains	1.6	50	1.8	100	4.5
Weighted average		116		156	

Source: From crop budgets prepared by the Central Planning Authority, Ministry of Agriculture.
Note: The N content in fresh matter is subject to considerable variation. The figures in the table are therefore indicative and intended only for comparison between the two periods. Nitrogen in plant parts not harvested has not been considered.

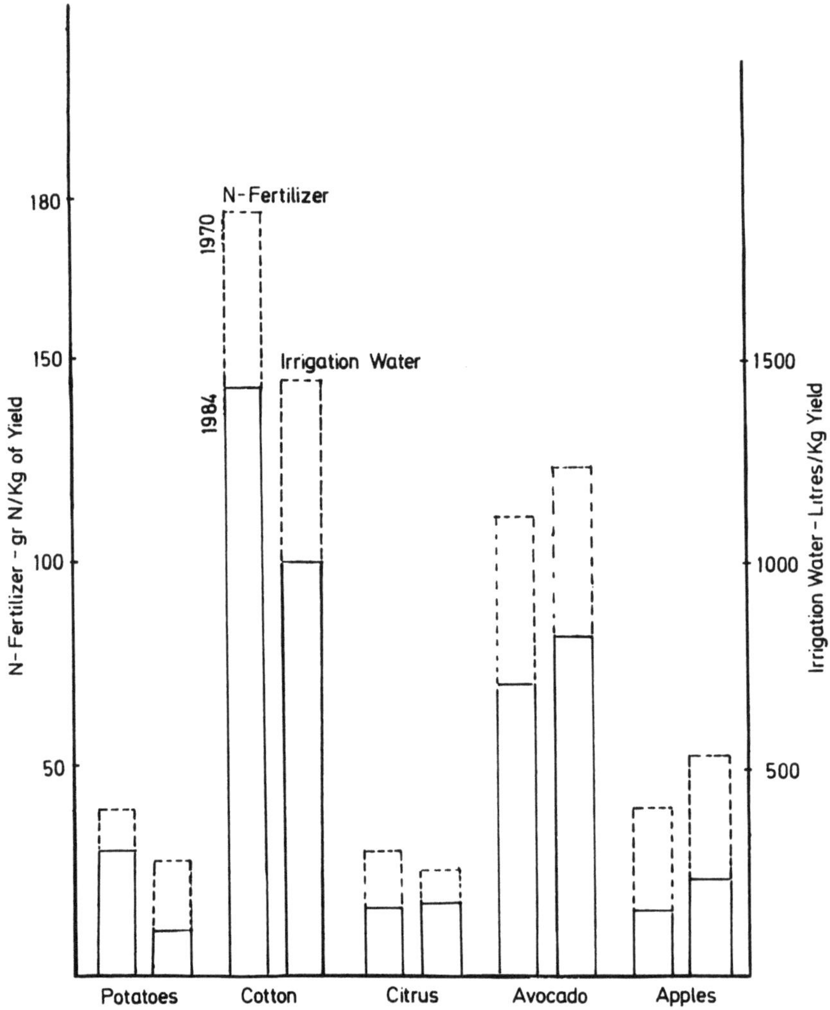

Fig. 4. N-fertilizer and irrigation water application rates for major crops, 1970 and 1984

3.3 Irrigation and Fertilizer Application Techniques

New Irrigation Techniques. In the past few decades, irrigation techniques have advanced from gravitation and sprinkler systems to drip, micro-sprinklers and mini-sprayers, and mechanized moving-line systems. In addition, computerized automated controls have been introduced to regulate irrigation at both the plot level and for the whole irrigation network. These new techniques allow control of the timing and dosage of various chemicals during the irrigation cycle (Beth 1981).

Such new irrigation techniques are widely employed in Israel, with the drip system already covering more than 80 000 ha, or about one-third of the total irrigated land (Segal 1986). These new irrigation techniques are characterized by partial wetting of the root zone, thus minimizing leaching and increasing the uptake efficiency by crops (Klein 1983).

Fertigation. Drip irrigation and similar techniques permit the administration of fertilizers with irrigation water. This technique, known as fertigation, ensures that the application of water and fertilizer is restricted to the root zone. Vegetative growth can thus be effectively controlled by limiting the water dose and by adjusting the quantity and timing of fertilizer application to the vegetative cycle (Klein 1983).

Dose Splitting. Splitting the nitrogen application increases fertilizer use efficiency by reducing loss through leaching and volatization. Gerwing et al. (1979) reported that the proportion of N-fertilizer taken up by the plants increased from 30.4% to 52.1% when it was applied in several small doses over the growing season as compared to a one-time application prior to planting. In Israel, the transition to fertigation has increased the number of applications by to 2–3 in maize and up to 12–17 times in cotton (Beth 1981).

3.4 N-Fertilizer Efficiency

The combined effect of all the various improvements made in farm husbandry practices, and especially in irrigation and fertilization techniques, has been to significantly reduce nitrogen and irrigation water application rates per unit weight yield (Fig. 4). In cotton, which is cultivated on about 30% of the total irrigated area, the nitrogen application rate was reduced from about 180 g kg^{-1} yield in 1970 to about 140 g kg^{-1} yield in 1984. Within the same period the irrigation dose was reduced from 1400 yield to about 1000 l kg^{-1} yield. Similarly, in the citrus groves, which cover 20% of the irrigated area, the N application rate was reduced from 30 to 15 g kg^{-1} and the irrigation dose from 250 to 170 l kg^{-1} between 1970 and 1984.

3.5 N-Removal in Harvested Crops and N-Residues in Soil

Israel Country-Wide Data. Based on the nitrogen content in plants, N application rates, and average yields, the amounts of N removed by harvested crops were estimated for the entire cultivated area of Israel, as shown in Table 3.

The summary data indicate that N residues in the cultivated areas of Israel, calculated as the difference between N applied and N uptake by harvested crops, have declined from about 22 000 tons in 1970 to about 18 000 tons in 1984, indicating an average drop in N residues from 65 kg N ha^{-1} in 1970 to 51 kg N ha^{-1}

Table 3. N-fertilizer application, N-uptake, and N-residues in soils in the cultivated area of Israel, 1970 and 1984

Year Crop	Area (1000 ha)	N-fertilizer (ton)				
		Application (kg ha^{-1})	Total applied	Total uptake	Total residue	Efficiency (%)
1970						
Fruit trees	70.7	189	13 367	2408	10 954	18
Field crops						
Cotton	55.5	110	6105	4396	1709	72
Other crops	52.1	225	11 718	5865	5858	50
Non-irrigated						
Various grains	170.9	50	8545	4922	3623	58
Total	342.9	116	39 695	17 289	22 109	44
1984						
Fruit trees	75.9	155	11 759	3308	8451	28
Field crops						
Cotton	61.1	130	7943	6599	1344	83
Other crops	78.2	277	21 694	17 262	4432	79
Non-irrigated						
Various grains	142.7	100	14 270	10 274	3996	72
Total	357.9	156	55 666	37 443	18 223	67

Sources: Data for 1984 compiled from data given in the Statistical Abstract of Israel (1985). Data for 1970 compiled from Saliternick et al. (1972).

in 1984. The corresponding figures for fruit tree crops, field crops, and non-irrigated crops are also given.

Coastal Plain Region. The same analysis was repeated for the most intensively cultivated area, the Coastal Plain Region, overlying the coastal aquifer which is one of Israel's major water resources. The results indicate a similar increase in fertilizer use efficiency and a decline in N residues in the soil from about 9000 tons in 1970 to about 6000 tons in 1984. This indicates an average drop in N residues from 102 kg N ha^{-1} in 1970 to 70 kg N ha^{-1} in 1984 (Table 4).

3.6 Relative Contribution of N-Fertilizer to the Pollution Load

The application of nitrogenous chemical fertilizers and manure to cultivated land is only one of several sources contributing to the nitrogen load in soil (Waldman and Shevah 1984). Other sources include grazing animal manure, sewage effluents, solid waste and rainfall. Comparison of these data (Table 5) obtained by Saliternick et al. for all sources in 1970 with those obtained for 1984 indicates that in the Coastal Plain Region the relative contribution of N-fertilizers to the

Table 4. N-fertilizer application, N-uptake, and N-residues in soils in the cultivated area of the Coastal Plain Region, 1970 and 1984

Year Crop	Area (1000 ha)	N-fertilizer (ton)				
		Application ($kg\ ha^{-1}$)	Total applied	Total uptake	Total residue	Efficiency (%)
1970						
Fruit trees	39.5	189	7641	1580	6061	21
Field crops						
Cotton	13.9	110	1529	1109	420	73
Other crops	20.2	225	4704	2321	2383	48
Non-irrigated						
Various grains	16.6	50	830	478	352	58
Total	90.2	163	14 704	5488	9216	37
1984						
Fruit trees	38.8	155	6250	2062	4188	33
Field crops						
Cotton	15.0	130	1950	1620	330	83
Other crops	23.0	277	6334	5072	1262	79
Non-irrigated						
Various grains	13.1	100	1310	943	367	72
Total	89.9	176	15 844	9677	6147	61

Sources: Data for 1984 compiled from data given in the Statistical Abstract of Israel (1985). Data for 1970 compiled from Saliternick et al. (1972).

Table 5. Sources of nitrogen pollution, 1970 and 1984

Area	Source	1970		1984	
		(1000 t)	(%)	(1000 t)	(%)
Coastal Plain Region	N-fertilizer	9.2	48.2	6.1	34.3
	Manure	2.3	12.0	3.1	17.4
	Sewage	3.8	20.0	4.1	23.0
	Solid waste	0.5	2.6	0.6	3.4
	Irrigation water	1.8	9.4	2.2	12.4
	Rainfall	1.5	7.8	1.7	9.5
	Total	19.1	100.0	17.8	100.0
Total Israel	N-fertilizer	22.1	48.7	18.2	38.6
	Manure	7.2	15.9	9.7	20.6
	Sewage	6.7	14.8	8.5	18.1
	Solid waste	0.8	1.8	1.1	2.3
	Irrigation water	2.5	5.5	3.0	6.4
	Rainfall	6.0	13.3	6.6	14.0
	Total	45.3	100.0	47.1	100.0

Source: Compilation from Saliternick et al. (1972), updated to 1984.

nitrogen load in soil has dropped from 48% in 1970 to 34% in 1984, while the corresponding figures for the country as a whole are 49 and 39%. On the other hand, the relative contribution of other polluting sources has increased with population growth.

4 Discussion

The direct relationship between N-application practice and excessive nitrate concentration in the soil profile is extensively documented. Nightingale (1972) found an agreement between observed groundwater NO_3-levels, soil NO_3-concentration, and N-fertilizer management practices, while Pratt et al. (1972) and Freid et al. (1976) concluded that the amount of N-NO_3-available for leaching is strongly related to nitrogen application practices. If N-fertilizer application did not exceed crop needs, no NO_3-N would be available for leaching.

Experiments carried out at ICI's Jeallotts Hill Research Station in the United Kingdom between 1981 and 1986 indicate that except when fertilizer applications are excessive, modification of agricultural practices, by sowing cereals to take up residual N in the soil, avoiding deep ploughing, and careful adjustment of fertilizer application could contribute to the alleviation of the nitrate problem. These research findings and experimental results are substantiated by Israel's practical experience: significant increase in fertilizer application efficiency can be achieved even when usage is increasing, if a transition is made from simple fertilizers to mixed NPK fertilizers formulated to suit specific crop and soil requirements. In addition, timely and controlled application throughout the vegetative cycle can prevent seepage of excess irrigation water and fertilizer to depths below the root zone (Beth 1981, Klein 1983).

Improvements in nitrogen application and irrigation practices have resulted in higher N-uptake by the plants. Combined with other improvements this has resulted in a substantial increased in yields of wheat, potatoes, maize, and other major crops (Table 2). In cotton the growing cycle was shortened, allowing a double cropping system of cotton followed by a "cleaning" crop (wheat) within the same year. This resulted in a further increase in N-uptake by plants and less residue in soil (Bielorai 1986).

From the environmental point of view, the increasing efficiency of N-fertilizers results in less N-fertilizer accumulated in cultivated land. In the case of Israel, the N-fertilizer contribution to the total pollution load has been reduced from about 50% in 1970 to about 40% in 1984, while in the ecologically sensitive Coastal Plain area the decline was even greater, from 48 to 34%. However, it is to be noted that because of the increased contribution of other polluting sources, in particular manure, sewage effluents and other solid wastes, the total amount of N reaching the soil has increased slightly.

These data presented in this chapter indicate that the common linkage between fertilizers and environmental pollution is not so obvious as it is generally assumed and sources of nitrogen pollution other than fertilizer should be investigated further. Initial findings would seem to indicate that pollution could be further reduced by better handling of non-agricultural wastes as well as by the continued application of the improved fertilizer management techniques already adopted by the farming community.

Acknowledgements. The authors wish to acknowledge the contribution of Uzi Montag of Haifa Chemicals Ltd. and Connie Wilsack for their assistance in the provision of data and for helpful comments.

References

Beth F (1981) Fertilization. Water & irrigation review. Water Workers Assoc Isr Jan:16-17

Bielorai H (1986) Irrigation systems and practices. In: Advances in irrigation and water supply. 4th Int Conf on Irrigation. Tel-Aviv, Israel Water Workers Assoc Publ pp 18-21

Conrad J (1986) Nitrates in ground and drinking water: analysis of policies and regulations. Sci Total Environ 51:209-225

Freid M, Tanji K, Van De Pol RM (1976) Simplified long term concept for evaluating leaching of nitrogen from agricultural land. J Environ Qual 5:197-200

Gerwing JR, Caldwell AC, Goodroad LL (1979) Fertilizer nitrogen distribution under irrigation between soil, plant and aquifer. J Environ Qual 8(3):281-284

Israel Central Bureau of Statistics (1985) Statistical abstract of Israel, Jerusalem, No. 36

Jeallot's Hill Research Station Guides (1981-1986) The fate of nitrogen investigations. ICI, Bracknell, Berks, UK

Jollans JL (1985) Fertilizers in UK farming. Centre Agric Strat, CAS Rep 9, Reading Univ

Klein Y (1983) Irrigation and fertilization of fruit trees by drip irrigation. Water and irrigation review. J Water Workers Assoc Isr Jan:8-11

Marom D (1986) Calculation of fertilizer application cost for various alternatives. Hassadeh 66 (8):1540-1542 (in Hebrew)

Nightingale MI (1972) Nitrates in soil and ground water beneath irrigated and fertilized crops. Soil Sci 114:300-311

Nitrogen (1986) Nitrate pollution: implications for fertilizer use. Nitrogen 162:22-25, British Sulphur Corp. London

Pratt PF, Jones WW, Hunsaker WF (1972) Nitrates in deep soil profiles in relation to fertilizer rates and leaching volume. J Environ Qual 1:97-102

Ronen DM, Magaritz M (1985) High concentration of solutes at the upper part of the saturated zone of a deep aquifer under sewage irrigated land. J Hydrol 80:311-323

Ronen D, Kanfi Y, Magaritz M (1984) Nitrogen presence in groundwater as affected by the unsaturated zone. In: Yaron B et al (eds) Pollutants in porous media. Springer, Berlin Heidelberg New York, pp 223-236

Saliternick C, Kahanovitz Y, Shevah Y (1972) Sources of pollution of groundwater by nitrogen compounds. Tahal Tel-Aviv, (in Hebrew)

Segal D (1986) Irrigation systems developments in Israel in the past decade. In: Advances in irrigation and water supply. 4th Int Conf on Irrigation. Tel-Aviv, Israel Water Workers Assoc Publ pp 35-39

Waldman M, Shevah Y (1984) Prevention of groundwater pollution on a national scale. In: Yaron B, Dagan G, Goldshmid J (eds) Pollutants in porous media. Springer, Berlin Heidelberg New York, pp 237-246

Water and Anion Transport in the Unsaturated Zone Traced by Environmental Tritium

H. Gvirtzman and M. Magaritz[1]

1 Introduction

Much information is available today about the dynamics of water and solute in porous media, and many theoretical explanations have been developed based on laboratory experiments (Nielsen et al. 1986). However, the application of these theories to natural systems is quite limited. Transport in the unsaturated zone, from the land surface down to the water table, differs significantly from transport in laboratory experiments using columns filled with various granules. It is different not only because of its scale and its boundary conditions, but also because of the complex structure of natural materials and the slow rate of processes taking place there. Consequently, in situ field measurements are necessary in order to understand these complex phenomena in a quantitative mode. This review summarizes the methodology of three field studies, in which water from the unsaturated zone was dated in a way that allows selection of a transport model best describing the transport mechanism in the studied medium.

One of the techniques for the quantitative study of natural conditions is the use of environmental tritium as a tracer. The principle is to date the water molecules along the sediment column according to their tritium content, using background tritium concentration data in the applied water. Since tritium is incorporated directly in the water molecule and has a half-life of 12.43 years, it can measure travel-time to any given depth lasting several tens of years. In addition, advantage has been taken of an atmospheric tritium pulse produced by thermonuclear tests in the 1950's and 1960's which entered rainwater. These rains infiltrated the land surface, percolated downward and created a sequence of water layers along the sediment column which serves as a rainfall record. The intention of the methodology described below was to improve the resolution of travel-time measurements and, thus, improve the conclusions concerning convection and dispersion processes in soil and sediment profiles. By improving the resolution, theories which had previously been studied only under laboratory conditions could be evaluated under natural field conditions.

[1]The Weizmann Institute of Science, Isotope Department, Rehovot 76-100, Israel

2 Methods

A method which is hereafter referred to as the "Rainfall–Irrigation Tritium Layering" was developed in the course of this work. In this method core samples are taken in the field and tritium content measurements are conducted on the extracted water. This method is based on the fact that in regions of Mediterranean climate, rainfall takes place only during winter seasons, and in agricultural areas, additional irrigation is applied during the summers. In areas in which irrigation water is supplied from low-level tritium groundwater, which significantly differs from high-level tritium rainwater, it is posible to recognize seasonal water layers in the unsaturated zone. It was found in this study that these layers are not mixed with each other, and alternating layers of high-tritium rainwater and low-tritium irrigation water, corresponding to the last tens of years, can be recognized to depths of tens of meters.

Research wells were drilled at three locations in Israel: (1) Glil-Yam in the Coastal Plain, (2) Omer in the northern Negev and (3) Hatzor in the southern Inner Coast. The samples were collected at 20- to 50-cm depth intervals by a spiral or auger driller without addition of water. For each depth interval, samples were collected separately for tritium analysis, chemical analysis and moisture content determinations.

Water for tritium analyses was extracted from each sample in a vacuum distillation system. The tritium concentration was analyzed using either a proportional counting system or a liquid scintillation counter. Tritium concentration is expressed in tritium units (TU), where 1 TU corresponds to 1 atom of tritium per 10^{18} hydrogen atoms. The gravimetric water content was determined by drying to a constant weight at 105°C. Anions were extracted with distilled water. Cl^-, NO_3^- and $SO_4^=$ concentrations were analyzed with an HPLC Ion Analyzer.

3 Results and Discussion

Three different transport phenomena were detected in the natural system and analyzed quantitatively using the "Rainfall–Irrigation Tritium Layering" method. These case studies are described below.

3.1 Simultaneous Transport of Water and Anions

The first profile was sampled in Glil-Yam in the unsaturated zone of the Coastal Plain Aquifer, in an area where precipitation is 600 mm/winter and irrigation (essentially sewage effluents) is 450 mm/summer (Gvirtzman et al. 1986). Sequential layers of high and low tritium concentration, corresponding to winter rains and summer irrigations, were detected in the sandy sediments unit (and not in the clay-loam unit due to the sampling resolution). Using these variations and background tritium concentration data in the applied water, a 20-year record of vertical water transport along 27 m was reconstructed (Fig. 1a, b). Comparison between the water quantity applied to the field and the water quantity found along the sediment column (for the corresponding years) showed that about 19% of the total input water percolated downward. Based on tritium peaks identification, the vertical water velocity was calculated to be 0.7 ± 0.05 m yr^{-1} in the clay-loam and 2.3 ± 0.2 m yr^{-1} in the sandy sediments.

The knowledge of the periodic application of solutes on the land surface (sewage effluents, irrigation and fertilization) was used for the investigation of their transport through the vadose zone. Sulphate and chloride peaks were detected along the profile, due to the large differences of these anions between the water sources. The anion peaks were identified and dated by comparing the measured profiles to the chronological agricultural activities. The peaks of $SO_4^=$ along the profile were correlated with the winter periods when the fertilizer, ammonium sulphate, was applied. Similarly, the Cl^- peaks were correlated with the summer seasons when sewage effluent irrigation was applied (Fig. 1c,d). The anion flow rate was determined by the dated peaks. It was found that in the sandy unit, the vertical velocities of Cl^- and $SO_4^=$ were the same as the velocity of the water. However, in the clay-loam unit, the vertical velocities of the anions were twice that of the water (1.35 ± 0.05 m yr^{-1}). This is clear evidence of the anion exclusion phenomenon under field conditions. Anions are repelled from negative clay surfaces by electrostatic forces and are concentrated in the center of the pores where water velocity is relatively fast. Schematic presentation of the unsaturated porous media is shown in Fig. 2, illustrating the anion-free water films (excluded volume) as well as the unexcluded volume. These field data fit and support the theoretical relationship between the water film thickness and the relative exclusion concentration proposed by Bresler (1973).

3.2 Mobile and Immobile Water Domains

The second profile was sampled in Omer in the northern Negev in a vineyard, in an area where precipitation is 200 mm/winter and irrigation is 650 mm/summer (Gvirtzman and Magaritz 1986). The unsaturated zone consists of loess sediments containing large quantities of salts. A 14-year record of vertical flow was reconstructed (Fig. 3a,b,). Alternate layers with high and low tritium concentrations, corresponding to winter rains and summer irrigations, were detected down to 8.5 m. Based on the peak separation and their dates, the vertical water velocity was calculated to be 0.66 ± 0.03 m yr^{-1} (by dividing each travel distance by the corresponding percolating time).

This tritium profile corresponds to a period during which the tritium content in the applied water was decreasing (1971–1983). However, the actual profile shows the opposite trend: an excess of tritium in the upper part of the section, and a deficiency of tritium in the lower part. This discrepancy led to interpretation of the tritium profile in terms of mobile and immobile water domains as the

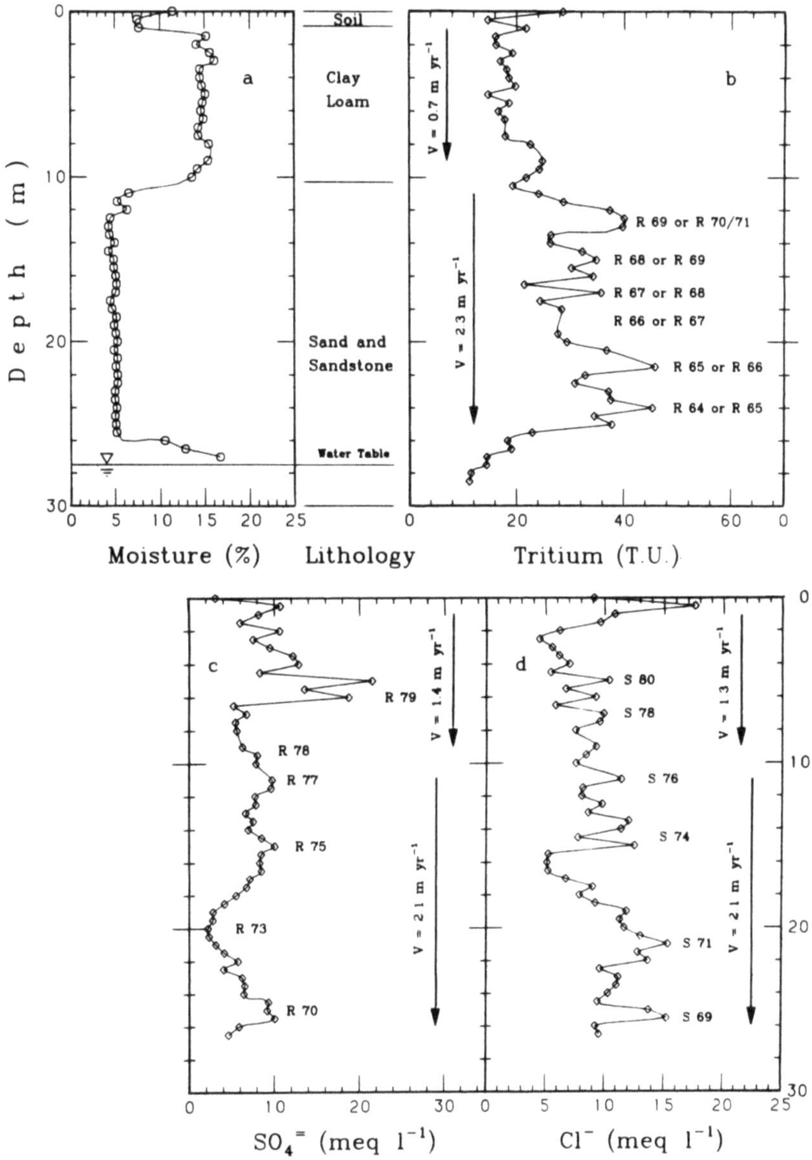

Fig. 1a-d. A profile of the Coastal Plain site, showing gravimetric water content **(a)**, tritium **(b)**, $SO_4^=$ **(c)** and Cl^- **(d)** distributions. Year and season (R = winter rain, S = summer sewage irrigation) of peaks and velocities (V) are indicated

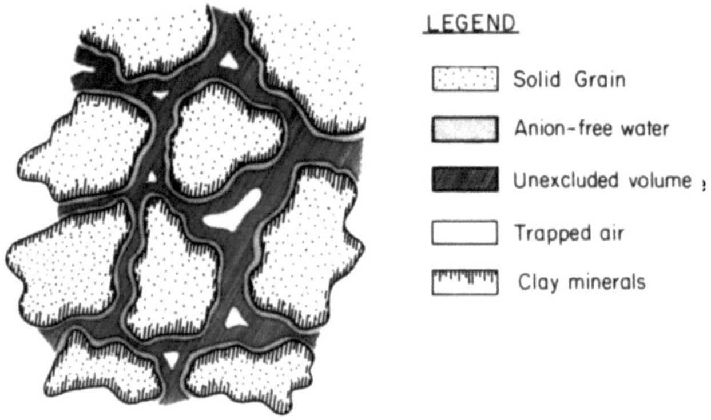

Fig. 2. A conceptual model of the unsaturated zone of the Coastal Plain site, which contains anion-free and unexcluded water films, exhibiting the anion exclusion phenomena

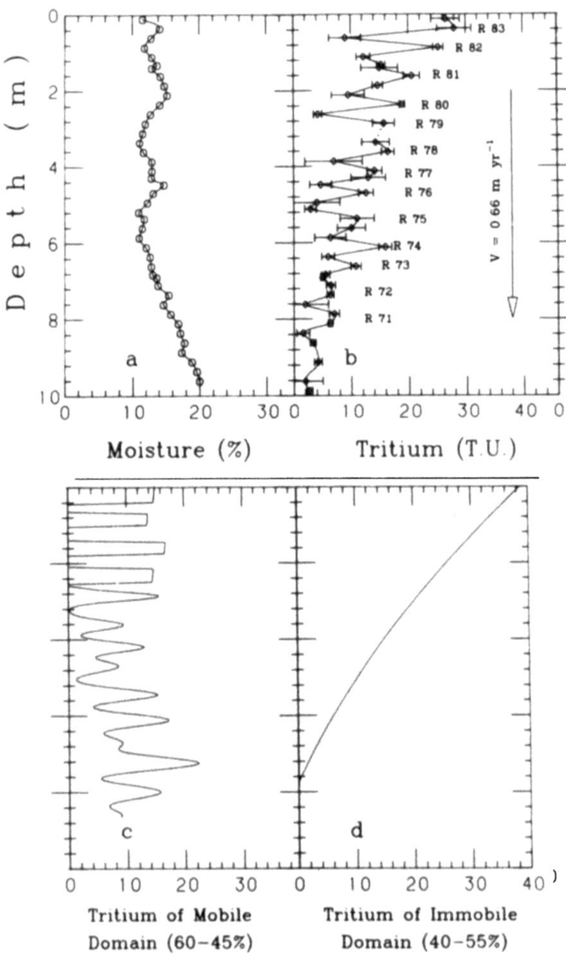

Fig. 3a-d. A profile of the northern Negev site, showing gravimetric water content **(a)**, measured total tritium **(b)**, the theoretical tritium profile of the mobile water domain **(c)** and the theoretical tritium profile of the immobile water domain **(d)**. *Error bars* indicate analytical uncertainty. Peak dates (*R* = rain) and velocity (*V*) are indicated

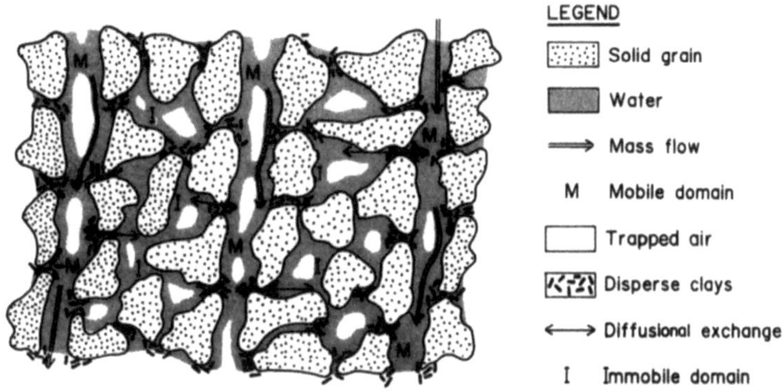

Fig. 4. A conceptual model of the unsaturated zone of the northern Negev site, which consists of mobile and immobile water domains. Gels of disperse clay minerals separate the two domains. *Flow lines* in the mobile domain and exchange directions between the two domains are indicated

extracted water in the laboratory is a combination of mobile and immobile waters (Fig. 3c,d). It has been hypothesized that the mobile domain was represented by percolating fresh water of both rain and irrigation, and the immobile one was represented by isolated, fossil, saline water pockets surrounded by dispersed clay minerals. Schematic presentation of the unsaturated porous medium is shown in Fig. 4.

The large amount of salts found in the profile (Magaritz et al. 1988) could not be related to agricultural activities (unlike the previous case). It was found that these salts originated from dry and wet fallout during the last 10 000 years. Agriculture in the investigated field was initiated in 1957 and the original salty solution has since started to be leached. However, it seems that to date the soil solution is still in a transitional stage. The solute concentration difference between the original saline water and the percolating fresh water caused dispersion of clay minerals at the interface between the two water types, thereby clogging several pores and passages. This was the reason for the formation of the immobile water domain which was identified from the tritium data.

The transport of the mobile water is described by a convective-dispersive flow which includes simultaneous diffusion-controlled mass exchange with the immobile water. The equations suggested by Van Genuchten and Wierenga (1977), with an extension for radioactive decay, were applied to this case study:

$$\frac{\partial C_m}{\partial t} = D_m \frac{\partial^2 C_m}{\partial x^2} - \frac{q}{\phi_m \theta} \frac{\partial C_m}{\partial x} - \alpha(C_m - C_{im}) - \lambda C_m$$

$$\frac{\partial C_{im}}{\partial t} = \frac{\phi_m}{1 - \phi_m} \alpha(C_m - C_{im}) - \lambda C_{im}$$

where:

$$\varphi_m = \frac{\theta_m}{\theta_m + \theta_{im}} = \frac{\theta_m}{\theta},$$

where C_m and C_{im} are the solute concentration (M L^{-3}) in the mobile and immobile water domains respectively, D_m is the dispersion coefficient in the mobile domain (L^2 T^{-1}), θ is the total volumetric water content (L^3 L^{-3}), q is the water flux (L T^{-1}), θ_m and θ_{im} are the water content (L^3 L^{-3}) of the porous medium in the mobile and immobile water domains respectively, φ_m is the fraction of the water which is mobile, α is the exchange rate coefficient of solute between the mobile and immobile domains (T^{-1}), and λ is the rate coefficient of radioactive decay (T^{-1}), t is time (T), and x is distance (L). These questions, together with the relevant initial and boundary conditions, were solved numerically. Using an optimization procedure, the best fit parameters were obtained. The percentage of the total water which is immobile was found to be about 40% and the mass exchange rate coefficient was found to be about 0.1 yr^{-1}.

3.3 Matrix and Fissure Water Transport

The third profile was sampled in Hatzor in the southern Inner Coast beneath an eucalyptus forest, in an area where precipitation is 500 mm/winter (Gvirtzman et al. 1987). A 35-year record of vertical flow, along 40-m calcareous sandstone sediments, was reconstructed according to the tritium concentration profile (Fig. 5). The increase in tritium concentration (30 to 90 TU) along the profile down to

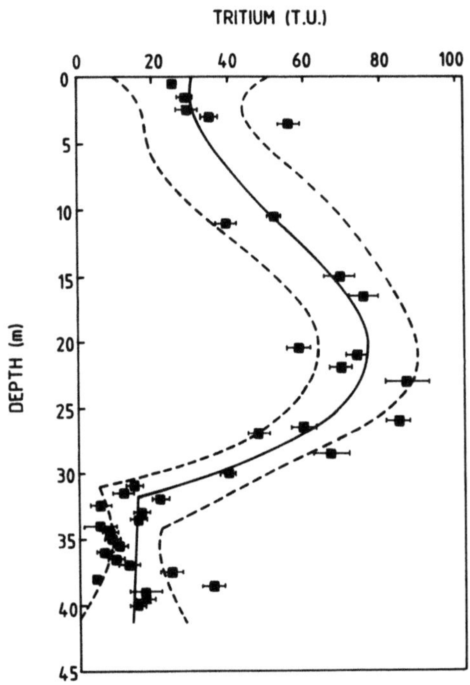

Fig. 5. A profile of the southern Inner Coast site, showing tritium concentration versus depth. *Error bars* indicate analytical uncertainty. The *solid lines* represent statistical regressions, and *dashed lines* represent the 95% confidence limits

30 m depth (where there is a sharp decrease in TU) represents the input of tritium associated with rainwater which fell in the period 1957–1985, while the relatively low concentrations (5 to 30 TU) below 30 m depth are probably pre-1957 rainwater. However, the tritium below 30 m depth are higher than the tritium in the rains of the pre-bomb period; thus, an additional mechanism, different from piston flow, is operating. The obtained tritium profile was explained in terms of matrix and fissure flow. Schematic presentation of the porous media is shown in Fig. 6. Using mass balance calculations, it was found that on the average, 40 mm yr^{-1} (8% of the yearly rain) percolated downward through the matrix pores at a velocity of 1.1 m yr^{-1}. An additional amount of more than 20 mm yr^{-1} (more than 4% of the yearly rain) percolated rapidly through the fissure network. These field data fit and support the model proposed by Wang and Narasimham (1985), which suggests that the bulk of the water movement under unsaturated conditions occurs through interconnected pores within the matrix.

Three different types of transport phenomena, operating in three different structures of porous media, were detected in the unsaturated zone. They were identified using the "Tritium Layering" method. Each flow mechanism was identified by its signature on the tritium concentration profile according to the similarity and/or discrepancy of the measured tritium profile and the theoretical profile expected from the simplest convective-dispersive model. The above three cases had the following signatures: In the first case, the similarity between the actual to the theoretical profile, i.e. the general trend of both increasing tritium content with depth, suggested that the simplest convective-dispersive flow is operating. The opposite trend of decreasing tritium concentration with depth, measured in the second case, suggested the existence of another component (i.e. immobile domain) along the whole profile. In the third case, a discrepancy was found at the bottom of the profile, suggesting the existence of another component of fast flow in the medium.

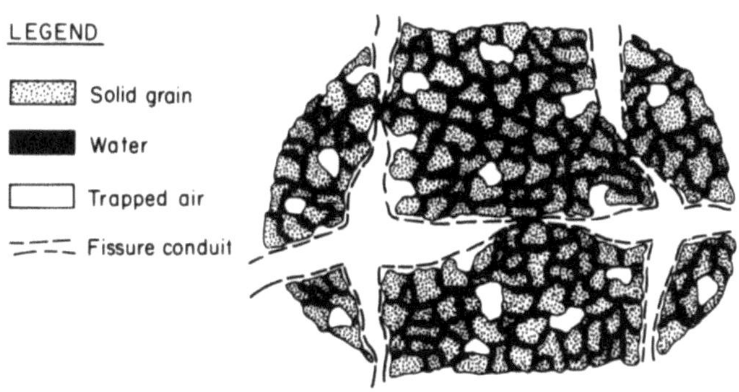

Fig. 6. A conceptual model of the unsaturated zone of the Inner Coast site, which contains fissure and matrix pores

4 Conclusions

Analysis of water transport mechanisms in a quantitative mode in natural systems represents a major contribution in understanding groundwater pollution processes and protecting water resources. The main obstacle in modelling water transport is the selection of the most appropriate model for the studied medium. The "Rainfall – Irrigation Tritium Layering" method helps to distinguish between the different transport phenomena under field conditions; thus, the selection of the most appropriate model can be achieved. Furthermore, several hydrological parameters might be determined using this method. However, it should be emphasized that the above method is restricted to fields receiving two sources of water (or more) which differ in their tritium concentration (or in any other chemical or isotopical tracer). In addition, reliable historical information concerning the applied water and solute is of critical importance.

References

Bresler E (1973) Anion exclusion and coupling effects in nonsteady transport through unsaturated soils: 1. Theory. Soil Sci Soc Am Proc 37:663–669

Gvirtzman H, Magaritz M (1986) Investigation of water movement through the unsaturated zone under irrigated area using environmental tritium. Water Resour Res 22:635–642

Gvirtzman H, Ronen D, Magaritz M (1986) Anion exclusion during transport through the unsaturated zone. J Hydrol 87:267–283

Gvirtzman H, Magaritz M, Kanfi Y, Carmi I (1988) Matrix and fissure water transport through unsaturated calcareous sandstone. Transport Porous Media 3:343–356

Magaritz M, Gvirtzman H, Nadler A (1988) Salt accumulation in the loessial sequence in the Be'er Sheva basin, Israel. Environ Geol Water Sci 11:27–33

Nielsen DR, Van Genuchten M Th, Biggar JW (1986) Water flow and solute transport processes in the unsaturated zone. Water Resour Res 22:89S–108S

Van Genuchten M Th, Wierenga PJ (1977) Mass transfer studies in sorbing porous media: 2, Experimental evaluation with tritium. Soil Sci Soc Am J 41:272–285

Wang JSW, Narasimhan TN (1985) Hydrologic mechanisms governing fluid flow in a partially saturated, fractured, porous media. Water Resour Res 21:1861–1874

Subject Index

MIX
Papier aus verantwortungsvollen Quellen
Paper from responsible sources
FSC® C105338
FSC
www.fsc.org

If you have any concerns about our products,
you can contact us on
ProductSafety@springernature.com

In case Publisher is established outside the EU,
the EU authorized representative is:
Springer Nature Customer Service Center GmbH
Europaplatz 3, 69115 Heidelberg, Germany

Printed by Libri Plureos GmbH
in Hamburg, Germany